Almost all the many recent studies on chaos have been concerned with classical systems. This book, however, is one of the first to deal with quantum chaos, the natural progression from such classical systems.

In this book the author deals with three major issues in quantum chaos. First, quantum mechanics is applied to both bounded and open systems exhibiting classical chaos. Potential problems involving quantum chaos are revealed in diverse areas of solid-state science, and standard concepts such as diamagnetism, antiferromagnetism, spin waves, electrical conductance and so on, are shown in a new light through quantum chaos. Second, adiabatic-ansatz eigenvalue problems are shown to yield a new paradigm of nonlinear dynamics, closing the gap between the greatly different theories of solitons and random matrices. Finally, the author provides a clue to how quantum mechanics may be impoved so as to accommodate temporal chaos.

This up-to-date book will be of value to researchers and graduate students in physics and mathematics studying chaos, nonlinear dynamics, quantum mechanics and solid-state science.

Quantum chaos

Cambridge Nonlinear Science Series 3

Titles in this series

1 Weak chaos and quasi-regular patterns
G. M. Zaslavsky, R. Z. Sagdeev, D. A. Usikov and A. A. Chernikov

2 Chaos, dynamics and fractals – an algorithmic approach to deterministic chaos
J. L. McCauley

3 Quantum chaos – a new paradigm of nonlinear dynamics
K. Nakamura

4 Thermodynamics of chaotic systems – an introduction
C. Beck and F. Schlögel

Quantum chaos

A new paradigm of nonlinear dynamics

Katsuhiro Nakamura

Osaka City University, Japan

CAMBRIDGE UNIVERSITY PRESS
Cambridge, New York, Melbourne, Madrid, Cape Town, Singapore, São Paulo, Delhi

Cambridge University Press
The Edinburgh Building, Cambridge CB2 8RU, UK

Published in the United States of America by Cambridge University Press, New York

www.cambridge.org
Information on this title: www.cambridge.org/9780521467469

First published 1993
First paperback edition 1994
Re-issued in this digitally printed version 2009

A catalogue record for this publication is available from the British Library

Library of Congress Cataloguing in Publication data

Nakamura, Katsuhiro, 1944–
Quantum chaos: a new paradigm of nonlinear dynamics/Katsuhiro Nakamura.
p. cm. – (Cambridge nonlinear science series; 3)
Includes bibliographical references and index.
ISBN 0-521-39249-7 (hc)
1. Quantum chaos. I. Title. II. Series.
QC174.17.C45N35 1993
530.1′2–dc20 92-7154 CIP

ISBN 978-0-521-39249-5 hardback
ISBN 978-0-521-46746-9 paperback

Contents

Preface

Despite the growing attention to chaos, its characteristic features, e.g. deterministic unpredictability, sensitivity to initial conditions and non-vanishing Kolmogorov–Sinai (KS) entropy are meaningful only in classical dynamical systems. Since contemporary science and high-technology have been founded on quantum mechanics, it is quite natural to proceed to considering the quantum-mechanical description of (classically) chaotic systems, which we shall call 'quantum chaos'. This subject has just begun to constitute an extremely challenging research field.

A consensus on a general framework, however, is not clear in this new field. Connections to classical concepts of chaos are not obvious. To speak more rigorously, the linear nature of the Schrödinger equation suppresses a chaotic diffusion of wavefunctions, leading to the vanishing of the KS entropy and of other characteristic exponents. Nonetheless, numerical evidence indicates that there are many interesting fundamental problems to be solved, e.g. irregular energy spectra and anomalous – though nonchaotic – quantum diffusions.

Theoretical tools borrowed from theories used in random systems are not always effective here. Wigner's level-spacing distribution in random matrix theory explains only a very limited aspect of the energy spectra of quantum chaos. The novel concept of localization of wavefunctions in disordered solid-state systems loses its significance when the localization length ($\propto$ (Planck's constant)$^{-2}$) is larger than the dimension of the Hilbert space. Therefore, this new field will require the creation of new ideas and concepts which have not been involved in conventional theories of random systems.

In this book I have tried to extract potential problems of quantum chaos in diverse branches of traditional solid-state physics. Well-known concepts such as diamagnetism, antiferromagnetism and spin waves will be shown in a new light in terms of quantum chaos. Readers will recognize quantum chaos to be a novel candidate for the disorder and noise observed in high-purity mesoscopic devices at very low temperatures.

No-one would doubt the validity of the Schrödinger–Feynman framework of quantum mechanics. Even Gutzwiller's semiclassical quantization of chaos remains within this framework. The present rapid progress in experiments on quantum dynamics in small molecules or quantum transport in mesoscopic devices may, however, possibly lead us to issues that cannot be explained within the established quantum mechanics. Then one should consider the invention of a new framework for quantum mechanics, heralding a new era of statistical mechanics and nonlinear dynamics. In chapters 5 and 6, I derive a remarkable dynamical system and its field-theoretical generalization that lies behind quantum chaos. However, the new paradigm of nonlinear dynamics does not merely imply these completely integrable systems. It means rather that a new framework of quantum mechanics be constructed in the near future so as to accommodate temporal chaos. This book is devoted to providing a clue to this great challenge.

The main part of this book is based on a series of my latest articles. So the content is extremely up-to-date and quite valuable for ambitious beginners in this field.

I am greatly indebted to many collaborators, in particular, C. Jeffries (University of California, Berkeley), P. Bryant (University of California, Santa Cruz) and P. Gaspard (Université Libre de Bruxelles). Further, I am grateful to many other people, including M. V. Berry, A. R. Bishop, O. Bohigas, H. Kuratsuji, M. Lakshmanan, H. J. Mikeska, Y. Nambu, S. A. Rice, A. Shudo, H. Thomas and H. Yamazaki for stimulating discussions. Daily discussions with K. Gohroku and K. Hirakawa at my Institute were also beneficial. Finally, I am extremely grateful to Mrs Chikako Nakamura for her endurance in the almost endless typing of the manuscript.

1
What are the problems?

1.1 Introduction

Quantum mechanics was born at the beginning of this century and is undoubtedly an indispensable guiding principle lying behind contemporary science and technology. A broad range of natural phenomena, such as ferromagnetism, superconductivity, superfluidity of helium and the quantum Hall effect can be explained by resorting to the quantum theory.

At the same time, however, the problems inherent in this theory such as particle–wave dualism, the subtle relationship between determinism and unpredictability encountered in measurement, have aroused profound and unsettled controversy among generations of physicists. For instance, the remark by Einstein, Podolsky and Rosen (1935) concerning a contradiction between the so-called Copenhagen interpretation and the principle of local causality is still close to our hearts (Bell, 1964; Wheeler and Zurek, 1983).

As we approach a new century, new kinds of problems about quantum mechanics are emerging which may demand revolutionary insights comparable to those at the beginning of this century. Among them two problems will be most intriguing. One is to unify quantum mechanics with general relativity or to construct a quantum-mechanically consistent theory of gravity. Tremendous efforts devoted towards the discovery of black holes are also promoting theoretical activity around this subject. Current theories, such as the Hawking radiation from black holes, will be incorporated into a new framework of 'quantum gravity' to be established in the near future. The other, and much more important problem is to search for the quantum-mechanical fingerprints of chaos or to develop the quantum mechanics of (classical) chaos. Our subject of 'quantum chaos' is, of course, concerned with this problem. Some, though not yet enough, experimental evidence of chaos in the microscopic cosmos has already been reported for hydrogen atoms in magnetic fields, ballistic electronic transport in mesoscopic semiconductor heterojunctions, spin-wave instabilities in small ferrites, and so on.

There have been many studies on chaos, fractals and nonlinear classical dynamics. Important theorems on the stability of regular phase space trajectories, for example, by Kolmogorov–Arnold–Moser (KAM), have been augmented by detailed numerical and analytic work on nonintegrable dynamical systems and maps. The quantum description of these classically nonintegrable systems (e.g., Hénon–Heiles, Sinai billiards and small molecules) now constitutes an extremely challenging research field. However, a consensus on a general framework is much less clear in the quantum case, where the existence, meaning, and nature of 'quantum chaos' are still strongly debated. Furthermore, with few exceptions, quantum dynamics has not been investigated, but rather the qualities of the distributions of eigenvalues and eigenfunctions have been considered. Therefore, connections to classical concepts of nonintegrability and chaos (intrinsically dynamic) are far from obvious.

At first sight, quantum mechanics in the form proposed by Schrödinger seems, without any substantial change, to be applicable to any system, whether it be classically regular or chaotic. One may, however, pose some natural questions. Can we actually see the quantum-mechanical counterpart of chaos despite the linear nature of the time-dependent Schrödinger equation? Why should the framework of quantum mechanics remain unaltered for (classically) chaotic systems? Will we want to change the definition of the de Broglie wavelength for a particle executing chaotic motion? I hope that readers will find some clues to the above questions in due course in the following chapters.

The concept of chaos is inherently relevant to classical dynamics. Therefore, the study of quantum mechanics in a semiclassical realm will be beneficial for us in giving an understanding of the effect of chaos on quantum mechanics. For the best analytical methodology to approach this issue, we should go back to the birth of quantum theory. For the moment, we shall briefly review the historical passage of quantum mechanics towards its ultimate establishment in 1925–6.

1.2 Adiabatic invariants in the history of quantum mechanics

The birth of quantum mechanics was brought about almost a century ago by great efforts to understand the laws describing radiation from a blackbody cavity and the specific heats of solids. While Planck first introduced energy quanta with a universal constant $h = 6.626 \times 10^{-27}$ erg s, the implicit idea of quantization can be traced to the Wien–Planck scaling law for the spectrum of blackbody radiation.

We now sketch Wien's formula (as reformulated by Ehrenfest) by using the concept of adiabatic invariants. We first consider a pendulum with mass m and string length l (see fig. 1.1). For small-amplitude oscillations, Newton's equation of motion is

$$ml\, d^2\theta/dt^2 = -mg\theta, \tag{1.1}$$

where g is the gravitational constant. The solution for (1.1) with an amplitude A and an initial phase δ is given by

$$\theta = A \cos(2\pi\nu t + \delta) \tag{1.2}$$

with the frequency

$$\nu = (1/2\pi)(g/l)^{1/2}. \tag{1.3}$$

The corresponding energy is given by

$$E = mglA^2/2. \tag{1.4}$$

If l is decreased by pulling the string upwards extremely slowly and smoothly, i.e., 'adiabatically', both the energy and frequency will be considerably increased. For a small adiabatic change $\delta l(<0)$, the work done on the string is

$$\delta W = -\langle F \rangle\, \delta l, \tag{1.5}$$

Fig. 1.1 Pendulum and adiabatic change of string length.

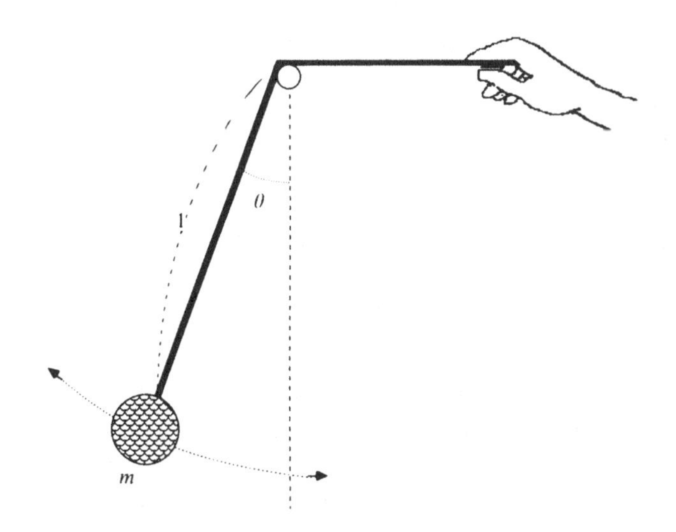

where $\langle F \rangle$ implies the time average of a reaction for the string tension $F = mg(1 - \theta^2/2) + ml(\mathrm{d}\theta/\mathrm{d}t)^2$. Using (1.2)–(1.4) in (1.5) and suppressing an oscillation-independent shift in the background potential energy, the increase in the pendulum energy is given by

$$\delta E = \delta W + mg\,\delta l = -El^{-1}\,\delta l/2. \tag{1.6}$$

The frequency increment, on the other hand, is

$$\delta \nu = -\nu l^{-1}\,\delta l/2. \tag{1.7}$$

From (1.6) and (1.7), we have $\delta E/E = \delta\nu/\nu$, giving that

$$E/\nu = \text{adiabatic invariant.} \tag{1.8}$$

This adiabatic invariance was first demonstrated by Einstein at the Solvay conference in 1911. It should be noted that the time average $\langle F \rangle$ in (1.5) is taken over the oscillation period of the pendulum, which is much smaller than the time required to cause the change $l \to l + \delta l$. If the period tends to infinity, we will inevitably fail to get adiabatic invariants like (1.8).

Coming back to the blackbody problem, energy resonators confined in a cavity with volume V at temperature T may be regarded as behaving as pendulums. So, for adiabatic contraction or expansion of the cavity, the resonators have adiabatic invariants as in (1.8). This adiabatic change, on the other hand, guarantees another invariant,

$$\nu_n/T = \text{const}, \tag{1.9}$$

which was shown by Wien using Stefan's law ($T \propto V^{-1/3}$ or $T \propto L^{-1}$) together with the resonator frequency $\nu_n = nc/L$. (L and c are the linear dimension of the cavity and the velocity of light, respectively; $n = 1, 2, 3, \ldots$.) In terms of the adiabatic invariants (1.8) and (1.9), Wien derived the scaling formula for the radiation spectrum which, in a modified form (Planck), is

$$E_n/\nu_n = \mathscr{F}(\nu_n/T). \tag{1.10}$$

Multiplying (1.10) by the state density of resonators $8\pi(V/c^3)\nu^2\,\mathrm{d}\nu$ we have

$$E(\nu)\,\mathrm{d}\nu = (8\pi V/c^3)\mathscr{F}(\nu/T)\nu^3\,\mathrm{d}\nu, \tag{1.11}$$

which turned out to show a remarkable agreement with experiments at all temperatures. Although Planck succeeded in determining the functional form $\mathscr{F}$ on the basis of his immortal idea of quanta, the Wien–Planck scaling behavior in (1.10) stood without any change throughout the quantum revolution of classical mechanics. Ehrenfest (1916) thereby concluded that in the adiabatic change, classical mechanics provides a key to the method of quantization, i.e.,

$$E/\nu = nh \qquad (n = 1, 2, 3, \ldots). \tag{1.12}$$

In general the adiabatic invariant on the left-hand side in (1.12) can be replaced by the action (rendering (1.12) into the Bohr–Sommerfeld rule for bounded systems):

$$J = \frac{1}{2\pi} \oint p \, \mathrm{d}q = n\hbar. \tag{1.13}$$

The integration in (1.13) is meaningful only for periodic orbits. The rule (1.13) was generalized to systems of $N > 1$ degrees of freedom as

$$J_k = \frac{1}{2\pi} \oint p_k \, \mathrm{d}q_k = n_k \hbar \qquad (k = 1, 2, \ldots, N), \tag{1.14}$$

if they were separable. To overcome the ambiguity in choosing separable variables in (1.14), Einstein (1917) proposed a more fundamental quantization rule:

$$J_k = \frac{1}{2\pi} \oint_{\Gamma_k} \sum_{l=1}^{N} p_l \, \mathrm{d}q_l = n_k \hbar \qquad (k = 1, 2, \ldots, M), \tag{1.15}$$

which is applicable also to nonseparable, but still integrable, systems. In (1.15) the total differential, $\sum_{l=1}^{N} p_l \, \mathrm{d}q_l$, is a canonical invariant; Γ_k denotes an irreducible (topologically distinct) closed contour in the N-dimensional invariant torus (see fig. 1.2 for $N = 2$). The number of constants of motion are represented by $M(=N)$.

Our overview so far, which is concerned with the so-called 'old' quantum theory or the semiclassical realm of quantum mechanics, indicates that the criteria of quantization are valid exclusively for classically periodic systems. For classically nonperiodic and chaotic

Fig. 1.2 Topologically independent paths on a two-dimensional torus.

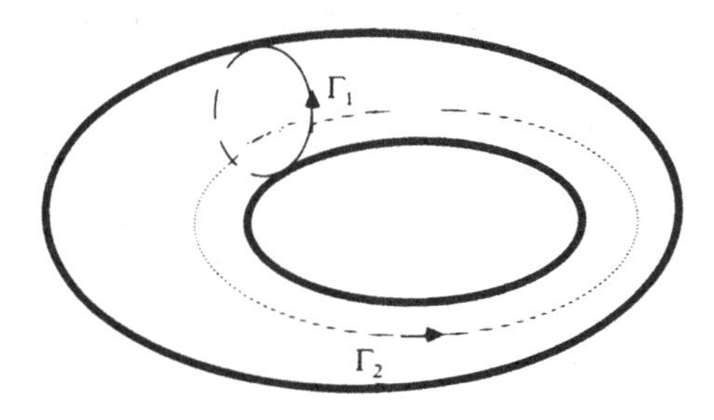

systems with $N > 1$ degrees of freedom, a rigorous correspondence in the adiabatic limit between quantum mechanics and classical mechanics will break down and the quantization procedure (1.15) will no longer be justified. In fact, in nonseparable systems, the invariant tori can be partially or fully destroyed, and thereby $M < N$. We find no way to proceed to full quantization. Fortunately these anxieties soon become insignificant.

Inspired by the revolutionary proposal by de Broglie concerning wave–particle dualism, Schrödinger established the wave-mechanics formalism of quantum mechanics, which was free from both the concepts of integral quantization and the problem of nonseparability. The Schrödinger (as well as the Heisenberg) formalism has since been considered as the ultimate quantum theory, because it has turned out not only to reproduce the rules in (1.15) for bounded systems (see the next section) but to explain all experiments concerning the microscopic world.

The problem of quantization of chaos has been incorporated into applying the quantum mechanics established in 1925–6 to classically chaotic systems. Other interesting quantum problems associated with chaos have also been reduced to seeing quantum-mechanical signatures of chaos.

However, this may not be the whole story. The progress of natural science is always heralded by unexpected results in experiments. While some experiments have been done on quantum systems which are classically chaotic, so far none of them seems to indicate the breakdown of quantum mechanics. Nevertheless, we cannot deny the possibility that thermal noise, extrinsic disorder, limited resolution in measurement, and so on, might have smeared out the fine quantum-mechanical structures originating from chaos. For the purpose of exposing a possible limitation of quantum mechanics and, further, to discover new criteria for the quantization of chaos, a modern analog of the blackbody cavity should be designed, which will convey precise information on quantum aspects of chaos in the microscopic world.

Promising experiments include those on quantum transport (conductance) in mesoscopic GaAs–AlGaAs heterojunctions with high purity at low temperatures (Beenakker and van Houten, 1991), where noninteracting electrons are executing two-dimensional ballistic motions extremely sensitive to the boundary of the system (i.e., the billiard boundary). Many examples of properties showing fluctuation due to the open nature of these systems have recently been reported. If careful experiments can succeed in demonstrating results incompatible with existing quantum theory, we should consider the possibility of inventing a new formalism of linear or nonlinear wave mechanics. At the same time, we should deal with anomalous commutation rules in the Heisenberg formalism of quantum mechanics and in quantum field theory.

1.3 Semiclassical quantization: Einstein, Brillouin and Keller versus Gutzwiller

In section 1.2, I mentioned that the endeavors to develop quantization conditions like (1.15) became insignificant following the discovery of quantum mechanics in 1925–6. As we have recognized, those quantization conditions can be traced back to the Wien–Planck scaling formula and are therefore guided by experiments. On the other hand, Schrödinger's quantum mechanics, which is also supported by an accumulation of experimental results on both closed and open systems, should include (1.15) at some appropriate limit. We illustrate this point below.

Considering a stationary problem, the time-independent Schrödinger equation is

$$H(\mathbf{q})\Psi_n(\mathbf{q}) = E_n\Psi_n(\mathbf{q}), \tag{1.16}$$

where $H(\mathbf{q}) = -(\hbar^2/2m)\sum_{j=1}^{N} \partial^2/\partial q_j^2 + V(q_1, q_2, \ldots, q_N)$ is the Hamiltonian for a particle with mass m moving in N dimensions. The corresponding Green function is defined by

$$G(\mathbf{q}'', \mathbf{q}'; E) = \sum_n \Psi_n^*(\mathbf{q}'')\Psi_n(\mathbf{q}')/(E - E_n). \tag{1.17}$$

Once the Green function has been obtained, we can easily find the eigenvalues E_n, spectral density $\rho(E)$ and eigenfunctions $\Psi_n(\mathbf{q})$. In fact,

$$\mathrm{Tr}\, G(E)\left(\equiv \int \mathrm{d}\mathbf{q}\, G(\mathbf{q}, \mathbf{q}; E)\right) = \sum_n \frac{1}{E - E_n}, \tag{1.18}$$

$$-\pi^{-1}\,\mathrm{Im}(\mathrm{Tr}\, G(E)) = \sum_n \delta(E - E_n)(\equiv \rho(E)), \tag{1.19}$$

$$\mathrm{Res}\,\{G(\mathbf{q}'', \mathbf{q}'; E = E_n)\} = \Psi_n^*(\mathbf{q}'')\Psi_n(\mathbf{q}'). \tag{1.20}$$

In (1.20), Res $\{\cdots\}$ implies a residue at the pole $E = E_n$. Since the Green function (1.17) is nothing but the Laplace transform of the time evolution propagator

$$K(\mathbf{q}'', \mathbf{q}'; t) \equiv \langle \mathbf{q}''|\mathrm{e}^{-\mathrm{i}Ht/\hbar}|\mathbf{q}'\rangle, \tag{1.21}$$

the problem of solving (1.16) is eventually reduced to obtaining K.

To see the connection with classical paths, Feynman's formalism of quantum mechanics (Feynman and Hibbs, 1965) is most advantageous, according to which K in (1.21) is given by

$$K(\mathbf{q}'', \mathbf{q}'; t) = \int_{\mathbf{q}(0)=\mathbf{q}'}^{\mathbf{q}(t)=\mathbf{q}''} \mathscr{D}[\mathbf{q}] \exp\{(\mathrm{i}/\hbar) W[\mathbf{q}]\}, \tag{1.22}$$

where W is a classical action functional expressed in terms of a classical Lagrangian $\mathfrak{L}$:

$$W[\mathbf{q}] = \int_0^t \mathfrak{L}(\mathbf{q}, \dot{\mathbf{q}})\, \mathrm{d}t. \tag{1.23}$$

Although (1.22) implies the sum of integrations over an innumerable number of paths, the great simplification in the calculation occurs in the semiclassical limit where the action W is much larger than $\hbar$. In this limit, we can take a stationary phase approximation in the neighborhood of

$$\delta W[\mathbf{q}] = 0, \tag{1.24}$$

having

$$K(\mathbf{q}'', \mathbf{q}'; t) = (2\pi \mathrm{i}\hbar)^{-N/2} \sum_\alpha |\Gamma_\alpha|^{1/2} \exp\{\mathrm{i}W_\alpha/\hbar - \mathrm{i}\mu_\alpha\}, \tag{1.25}$$

where α denotes the classical orbits satisfying Hamilton's principle, (1.24), with boundary conditions $\mathbf{q}(0) = \mathbf{q}'$ and $\mathbf{q}(t) = \mathbf{q}''$. The values W_α, μ_α and Γ_α are defined for each orbit α: the phase shift μ_α, called the Morse or Maslov index, comes from caustics (singularities of Γ_α). Using the number of singular points m_α between $\mathbf{q}'$ and $\mathbf{q}''$,

$$\mu_\alpha = m_\alpha \pi/2. \tag{1.26}$$

The inverse Jacobian, Γ_α, is given by $\Gamma_\alpha = \det\{-\partial^2 W_\alpha/\partial \mathbf{q}''\, \partial \mathbf{q}'\}$.

As noted above, (1.21), the Laplace transform of the propagator K with $\mathbf{q}'' = \mathbf{q}'$ yields the Green function. In particular, its trace is

$$\operatorname{Tr} G(E) = \int \mathrm{d}\mathbf{q} \left\{ -(\mathrm{i}/\hbar) \int \mathrm{d}t\, \mathrm{e}^{(\mathrm{i}/\hbar)Et} K(\mathbf{q}, \mathbf{q}; t) \right\}. \tag{1.27}$$

The way of evaluating (1.27) strongly depends on the integrability or nonintegrability of the underlying classical system.

In the completely integrable case, the phase space is occupied by invariant tori as in fig. 1.2 and adiabatic invariants given by the actions

$$S_k = \oint_{\Gamma_k} \mathbf{p} \cdot \mathrm{d}\mathbf{q} \qquad (k = 1, 2, \ldots, N) \tag{1.28}$$

are essential. In (1.28), Γ_k implies an irreducible closed contour in fig. 1.2. After integration with respect to t in (1.27), we transform from $\mathbf{p}, \mathbf{q}$ to action-angle variables. Then, for any topologically closed loop, the effective action can be written as a sum of the winding number (l_k) times the action S_k. Eventually we have (Berry and Tabor, 1977a; Zaslavsky, 1981)

$$\mathrm{Tr}\, G(E) \simeq V \sum_{l_1=0}^{\infty} \cdots \sum_{l_N=0}^{\infty} \prod_{k=1}^{N} \mathrm{e}^{\mathrm{i} l_k (S_k/\hbar - \mu_k)}$$

$$= V \prod_{k=1}^{N} (1 - \mathrm{e}^{\mathrm{i}(S_k/\hbar - \mu_k)})^{-1}, \qquad (1.29)$$

where μ_k is now the Morse–Maslov index for Γ_k and V is the volume of the N-dimensional torus characterized by $\{\Gamma_k\}$. Poles of (1.29) yield

$$\frac{1}{2\pi} \oint_{\Gamma_k} \mathbf{p}(E) \cdot \mathrm{d}\mathbf{q} = (n_k + m_k/4)\hbar \qquad (n_k = 0, 1, 2, \ldots). \qquad (1.30)$$

This is just Einstein's quantization rule improved so as to include the Morse–Maslov indices. This rule in the form (1.30), which was reached originally by Brillouin and Keller by noting the single-valuedness of semiclassical wavefunctions (Van Vleck, 1928), is referred to as the Einstein, Brillouin and Keller or EBK quantization rule (Keller, 1958; Percival, 1977). The semiclassical limit of quantum mechanics has thus turned out to reproduce the result of 'old' quantum theory, thereby unambiguously establishing a one-to-one correspondence between the invariant tori and quantum eigenvalues.

As we learnt in section 1.2, the rule (1.30) is by nature traceable to experimental evidence (i.e., the Wien–Planck scaling formula). It holds good only for completely integrable systems, in which the number of adiabatic invariants, i.e., constants of motion in involution, accords with the degree of freedom N. Gutzwiller, however, proceeded to look for an analogous correspondence between the semiclassical quantum 'irregular' spectra and chaotic orbits in nonintegrable systems. Noting the absence of invariant tori and retaining $\mathbf{p}, \mathbf{q}$ coordinates throughout, $\mathbf{q}$ integration in (1.27) in the stationary phase approximation leads to the claim that $\mathbf{p}'' = \mathbf{p}' = \mathbf{p}$ and necessitates a patching of all (unstable) periodic orbits in (1.27):

$$\mathrm{Tr}\, G(E) = \sum_{\alpha} \sum_{l=0}^{\infty} f_\alpha(E) \exp\{\mathrm{i} l [S_\alpha/\hbar - \mu_\alpha]\}, \qquad (1.31)$$

where

$$f_\alpha(E) = (i\hbar)^{-1}(2\pi i\hbar)^{-(N-1)/2} T_\alpha |\Gamma_\alpha^\perp|^{1/2}. \tag{1.32}$$

Here α denotes a primitive periodic orbit with energy E and l the number of its repetition; the action $S_\alpha(E)$ and period $T_\alpha(E)(= -\partial S_\alpha/\partial E)$ are for the orbit α; $\Gamma_\alpha^\perp$ is an exponent responsible for the transverse orbital stability:

$$\Gamma_\alpha^\perp = \det\left(-\partial^2 S_\alpha/\partial \mathbf{q}_\perp'' \, \partial \mathbf{q}_\perp'\right)|_{\mathbf{q}''=\mathbf{q}'=\mathbf{q}}. \tag{1.33}$$

From (1.31)–(1.33), we finally reach Gutzwiller's trace formula (Gutzwiller, 1967, 1971, 1990; Balian and Bloch, 1972, 1974):

$$\begin{aligned} \operatorname{Tr} G(E) - \operatorname{Tr} G_0(E) = (i\hbar)^{-1} \sum_\alpha T_\alpha \sum_{l=1}^{\infty} \{\det(M_\alpha^l - 1)\}^{-1/2} \\ \times \exp\{il(S_\alpha/\hbar - \mu_\alpha)\}. \end{aligned} \tag{1.34}$$

$\operatorname{Tr} G_0(E)$ comes from a contribution from zero-length orbits. M_α^l is a linearized Poincaré map describing the time evolution of transverse displacement from the orbit α and its Lyapunov exponent u_α depends on the type of fixed points. Typically, for homoclinic orbits with hyperbolic fixed points,

$$\det(M_\alpha^l - 1) = 4\sinh^2(lu_\alpha/2). \tag{1.35}$$

From (1.34) one may understand that semiclassical quantum eigenvalues are constructed through complicated interference between a set of periodic orbits. When applied to integrable systems, (1.34) recovers the EBK quantization conditions. In chaotic systems, however, no invariant torus exists and α covers arbitrarily-long periodic orbits. The number of orbits with a period less than $T(\gg)1$ is $N(T) \simeq \exp(h_{KS}T)/T$, i.e., exponentially proliferating. In the trace formula (1.34), the amplitude of terms with a given period $T(\gg)1$ is $A(T) \simeq T\exp(-h_{KS}T/2)$. Multiplying $A(T)$ by $N(T)$ gives a contribution $\exp(h_{KS}T/2)$, which brings about a serious problem of nonconvergence in (1.34). Further, finding all periodic orbits without missing any of them is a formidable task. Nevertheless, for some fully chaotic systems without any bifurcation, symbolic codings for all periodic orbits have been derived. In fact, using periodic orbits, thus organized, Gutzwiller (1982, 1990) applied (1.34) to the anisotropic Kepler problem, obtaining low-lying eigenvalues in good agreement with exact quantum eigenvalues.

Nevertheless, symbolic coding of periodic orbits in generic systems is much less obvious. Even if one could achieve this coding, one would then meet another difficulty, namely that of nonconvergence in the Gutzwiller series noted above. Although we shall come back to this subject in the final chapter, it is true that many scholars are struggling with immense difficulties and, nevertheless, remain far from the ultimate goal of obtaining individual eigenvalues accurately. In conclusion, we may safely assert that (1) patching of periodic orbits, as it stands, is not the most plausible method for semiclassical quantization of chaotic orbits; (2) we should try to get a hint of a thoroughly new semiclassical quantization rule by designing novel experiments comparable to those concerning blackbody radiation; and (3) we may even conceive the possibility of constructing a new framework for quantum mechanics based on the nonlinear Schrödinger equation, so long as the validity of Schrödinger's quantum mechanics for classically chaotic systems is not fully indisputable.

Despite its serious drawbacks, Gutzwiller's work has played an important role for the following reasons. (1) It has inspired the study of periodic-orbit scars in wavefunctions (Heller, 1984; Bogomolny, 1988). (2) Using the trace formula, the theory of spectral rigidity has been developed (Berry, 1985). Thus it seems that the semiclassical region is the most attractive target for our study to elucidate the classical and quantum correspondence in chaotic systems. We shall henceforth envisage many more striking discoveries in this area and, in the final chapter, think again about a new framework for quantum mechanics.

1.4 Avoided level crossing and quantum adiabatic phase

In the remaining part of this chapter, we sketch several key concepts required to understand the following chapters.

The major problem of quantum mechanics (for stationary states) is to solve the eigenvalue problems (1.16) for the Hamiltonian, which in general depends on parameters $\mathbf{R} = (R_1, R_2, \ldots, R_d)$ such as external electric and magnetic fields. The eigenvalues are characterized by a set of symmetry-induced quantum numbers. Within a desymmetrized manifold, no degeneracy can be seen except for the accidental ones (von Neumann and Wigner, 1929): if a single parameter were to be varied, we should see only level repulsions, e.g., avoided level crossings. Let us examine this problem by using a two-states model with a Hamiltonian $H(\mathbf{R})$. Suppose that at some point $\mathbf{R}_0$ two states $|\Psi_1^0\rangle$ and $|\Psi_2^0\rangle$ are degenerate with common energy E_0. The eigenvalue problem in the close neighborhood of $\mathbf{R}^0$ can be solved by degenerate perturbation theory. Using as diabatic basis $|\Psi_1^0\rangle$ and $|\Psi_2^0\rangle$, the matrix elements of $H(\mathbf{R})$ are given as

$$\begin{bmatrix} E_0 + H'_{11}(\mathbf{R}) & H'_{12}(\mathbf{R}) \\ H'_{21}(\mathbf{R}) & E_0 + H'_{22}(\mathbf{R}) \end{bmatrix} \tag{1.36}$$

with $H'_{ij}(\mathbf{R}) = \langle \Psi_i^0 | \{H(\mathbf{R}) - H(\mathbf{R}_0)\} | \Psi_j^0 \rangle$. The energy difference at $\mathbf{R}$ is then

$$\Delta E = \{(H'_{11}(\mathbf{R}) - H'_{22}(\mathbf{R}))^2 + 4|H'_{12}(\mathbf{R})|^2\}^{1/2}. \tag{1.37}$$

To meet the degeneracy, both

$$H'_{11}(\mathbf{R}) = H'_{22}(\mathbf{R}) \tag{1.38a}$$

and

$$|H'_{12}(\mathbf{R})| = 0 \tag{1.38b}$$

should be satisfied. The number $\varkappa$ of mutually independent equations in (1.38*a*) and (1.38*b*) is 2, 3 and 5, respectively, for real symmetric, complex Hermitian and quaternionic Hamiltonians. Hence, if $d < \varkappa$, we cannot see any degeneracy except for accidental ones. For $d = \varkappa$ we can have a set of isolated degenerate points. Finally, for $d > \varkappa$ the degeneracy is not isolated, lying on a $(d - \varkappa)$-dimensional manifold, e.g., lines, surfaces, etc. This situation has been extensively studied by Arnold (1978) in the context of the theory on a surface of the second order (quadric): the condition for degeneracy to occur is identical to that for a given ellipsoid to become the ellipsoid of revolution.

In fully chaotic systems, we have no integral of motion other than the total energy. The corresponding quantum systems are therefore desymmetrized from the outset. When a single nonintegrability parameter is changed, we see many avoided level crossings (Berry, 1981). In the semiclassical limit the level density is very large and these avoided crossings constitute a backbone of complicated energy spectra. Some aspects of these spectra, e.g., level spacing distribution at fixed parameter value, have proved to be well described by random matrix theory (Bohigas *et al.*, 1984; Bohigas and Giannoni, 1984). We shall come back to this point in chapter 5.

In case of $d \geq 2$, the energy spectra are depicted in the multi-dimensional $\mathbf{R}$ space with each eigenvalue constituting a hyper-energy surface. If $d \geq \varkappa$ and $d \geq 2$, then adjacent energy surfaces can be connected at isolated point degeneracy ($d = \varkappa$), line degeneracy ($d - \varkappa = 1$), and so on. Eigenstates are multi-valued in the parameter space $\mathbf{R}$, though of course they are single-valued in the configuration space. Let us consider a closed circuit C in $\mathbf{R}$ space. For an adiabatic transport along C, the eigenfunction acquires a nonintegrable geometric phase (Berry, 1984)

$$\Gamma_n(C) = \oint_C \mathbf{A}_n(\mathbf{R}) \cdot d\mathbf{R}, \tag{1.39}$$

where $\mathbf{A}_n(\mathbf{R})$ is a fictitious gauge potential given by

$$\mathbf{A}_n(\mathbf{R}) = \mathrm{i}\langle n(\mathbf{R})|\nabla_{\mathbf{R}}|n(\mathbf{R})\rangle. \tag{1.40}$$

The phase (1.39) is a manifestation of anholonomy (i.e., failure of some variables to return to their original values after a cyclic change of other variables) for the parallel transport of quantum eigenstates (unit vectors) in Hilbert space. In the field of quantum chemistry, vibration–rotation energies of molecules are analyzed in the Born–Oppenheimer approximation, whereby electronic quantum states are assumed to follow the change of nuclear coordinates adiabatically. Longuet-Higgins *et al.* (1958) noticed the geometric-phase-induced sign change in the electronic wavefunctions when nuclear coordinates are cycled. In the field of differential geometry, Darboux (1896) had already found a similar sign change in his analysis of 'umbilic points' of curved surfaces. Nevertheless, the universal gauge structure around level degeneracies was not revealed until Berry's work in 1984.

By applying Stokes' theorem, $\Gamma_n(C)$ in (1.39) can be rewritten as

$$\begin{aligned}\Gamma_n(C) &= \mathrm{i}\iint_C \mathrm{d}S\cdot(\nabla_{\mathbf{R}} \times \mathbf{A}_n(\mathbf{R}))\\ &= \iint_C \mathrm{d}S\cdot V_n(\mathbf{R}),\end{aligned} \tag{1.41}$$

where S denotes an area element in $\mathbf{R}$ space.

In (1.41), $V_n(\mathbf{R})$ is a fictitious magnetic field defined by

$$\begin{aligned}V_n(\mathbf{R}) &= \mathrm{Im}\sum_{m(\neq n)} (\nabla_{\mathbf{R}}\langle n|)|m\rangle \times \langle m|\nabla_{\mathbf{R}}|n\rangle\\ &= \mathrm{Im}\sum_{m(\neq n)} \frac{\langle n|\nabla_{\mathbf{R}}H(\mathbf{R})|m\rangle \times \langle m|\nabla_{\mathbf{R}}H(\mathbf{R})|n\rangle}{(E_n(\mathbf{R}) - E_m(\mathbf{R}))^2}\end{aligned} \tag{1.42}$$

The form (1.42) indicates that $V_n(\mathbf{R})$ is singular at degenerate points.

Let us again consider a two-states model for a complex Hermitian Hamiltonian $H(\mathbf{R})$ with $\mathbf{R} = (X, Y, Z)$. Note that $d = \varkappa = 3$ in this model. Suppose that a degeneracy occurs at $\mathbf{R} = \mathbf{R}_0$, and then H can be approximated around $\mathbf{R} = \mathbf{R}_0$ as

$$H(\mathbf{R}) = H(\mathbf{R}_0) + (\mathbf{R} - \mathbf{R}_0)\cdot\nabla_{\mathbf{R}}H(\mathbf{R})|_{\mathbf{R}=\mathbf{R}_0}. \tag{1.43}$$

Putting $\mathbf{R}_0 = 0$ and $H(\mathbf{R}_0) = 0$ without loss of generality, we can get a universal SU(2) Hamiltonian in terms of Pauli matrices σ as

$$H(\mathbf{R}) = \tfrac{1}{2}\sigma\cdot\mathbf{R}. \tag{1.44}$$

A pair of energy surfaces is represented by

$$E_{\pm}(\mathbf{R}) = \pm\tfrac{1}{2}(X^2 + Y^2 + Z^2)^{1/2}, \tag{1.45}$$

and they are conically intersected at the diabolical points (see fig. 1.3). Using $\nabla_{\mathbf{R}} H(\mathbf{R}) = \sigma/2$ and substituting (1.45) in (1.42), we have

$$V_{\pm} = \pm\mathbf{R}/2R^3. \tag{1.46}$$

This looks like a magnetic field induced by a Dirac monopole with strength $\frac{1}{2}$ located at $\mathbf{R} = 0$. The geometric phase is now given by

$$\Gamma_{\pm}(C) = \mp\Omega/2, \tag{1.47}$$

where Ω is the solid angle for the view of circuit C as seen from the degeneracy. When C encloses (with any plane circle) the degeneracy, $\Omega = \pm 2\pi$; otherwise, $\Omega = 0$. In the former case,

$$\exp\{\mathrm{i}\Gamma_{\pm}(C)\} = -1, \tag{1.48}$$

which changes the sign of the eigenstates. In the path-integral formalism, the geometric phase leads to a modification of the quantization rule, and gives rise to anomalous terms in the quantum-mechanical commutation relations (Kuratsuji and Iida, 1988; Shapere and Wilczek, 1989).

So long as a single nonintegrability parameter is changed, no Dirac monopole can be seen, but the resultant complicated energy spectra with many avoided crossings can be interpreted as a vertical cross-section of a landscape of piled sheets of energy surfaces in E–$\mathbf{R}$ space with $\mathbf{R}$ in d

Fig. 1.3 Conical intersection at diabolical (degenerate) point.

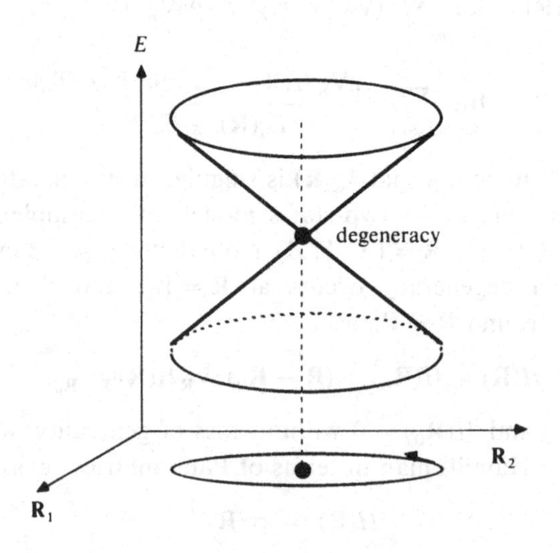

(≥ 2) dimensions (see fig. 1.4): these surfaces are mutually connected at diabolical conical points (Dirac's monopoles). In the semiclassical limit, there will be a dense monopole population with strong interactions between them. Derivation of the irregular spectra of quantum chaos by using the statistical mechanics of the underlying monopole gas or liquid, as a natural extension of the attempts in chapter 5, is a challenging problem.

Fig. 1.4 (*a*) Piled sheets of hyper-energy surfaces in case $d = 2$. Diabolical degenerate points and avoided crossings are indicated by $\ast$ and $\rightarrow$, respectively; (*b*) Vertical cross-section of (*a*).

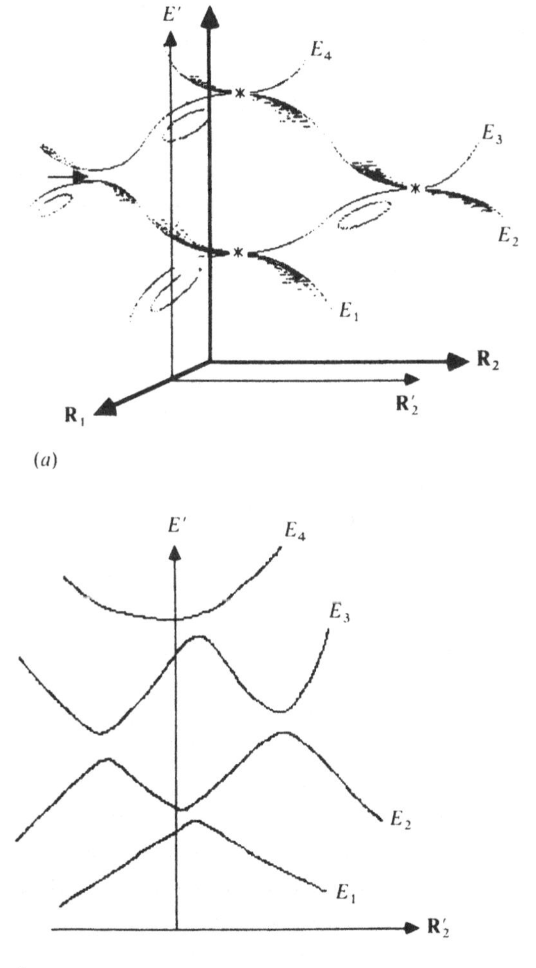

1.5 Quantum dynamics: ergodicity, recurrence and cross-over time

Other key concepts are derived from quantum dynamics. In classical dynamics, a working definition of chaos involves the exponential dependence of a final state on an initial state. This property, characterized in terms of the positive Lyapunov exponent and nonvanishing Kolmogorov–Sinai entropy, leads to ergodicity and mixing. The possible existence of corresponding properties in quantum mechanics has been debated with respect to the long-standing problem of quantum ergodicity, i.e., the problem of whether a long-time average of a given observable is equal to its statistical expectation.

Von Neumann (1929) was the first to give a dynamical foundation for quantum ergodicity. Since macroscopic observations are made within a small but finite time, they are always accompanied by a quantum uncertainty ΔE in energy. He therefore considered an $N(\gg 1)$-dimensional energy shell S of width ΔE within which a normalized state evolves. He investigated a mean square deviation (m.s.d.) of a macro-observable P, the projection operator onto an $M(\gg 1)$-dimensional subspace of S. The mean in m.s.d. here implies a time average. Using the occupation probability $P(t) = \langle \Psi(t)|P|\Psi(t)\rangle$ and the equal a priori probability $w = M/N$, the m.s.d. is defined by

$$\overline{(P(t) - w)^2} = \lim_{T\to\infty} T^{-1} \int_0^T \mathrm{d}t (P(t) - w)^2. \tag{1.49}$$

The quantum ergodicity for a micro-canonical ensemble would be proved if we could show that, irrespective of initial states, $\overline{P(t)} = w$. In practice, it is hard to give this proof. Instead, von Neumann obtained the inequality

$$\langle \overline{(P(t) - w)^2} \rangle_P / w^2 < 2(1 - w)/M \tag{1.50}$$

by assuming the energies in S to be both nondegenerate and nonresonant. The additional average over P in (1.50) is taken for mathematical convenience. Inequality (1.50) implies that a typical macro-observable P of large enough dimension ($M \gg 1$) almost always takes the statistical expectation, i.e., the a priori probability.

Von Neumann's assertion was later criticized by many groups. Bocchieri and Loinger (1958, 1959), in particular, concluded that the inequality in (1.50) is a consequence of the average over P and has nothing to do with quantum dynamics. The new inequality derived by them is

$$\langle \overline{(P(t) - w)^2} \rangle_\Psi / w < (1 - w)/M, \tag{1.51}$$

where the additional average is taken over initial states $\Psi(0)$ rather than P. (See Pechukas (1984) for a review.)

Despite their historical significance, however, the studies on quantum ergodicity described above have no practical significance. While in classical mechanics time-averaging over an infinite interval is essential to distinguish between ergodic and nonergodic motions, an analogous averaging in quantum mechanics smears out all dynamical information. In contrast to classical ergodic theory (e.g., Birkhoff, 1931), the most interesting aspect of quantum dynamics lies in its finite-time behavior rather than its averaging over an infinite time.

Studies in this direction were developed by Bocchieri and Loinger (1957) themselves and Hogg and Huberman (1982), who showed quantum recurrence phenomena, i.e., a quantum analog of Poincaré's recurrence theorem. By quantum recurrence, it is meant that both wavefunctions and energies reassemble themselves infinitely often in the course of time evolution.

Consider the time-dependent Schrödinger equation (Hogg and Huberman, 1982)

$$i\hbar\, \partial\Psi/\partial t = H(t)\Psi \tag{1.52}$$

with an initial condition $\Psi_0 = \Psi(t_0)$. The periodicity $H(t + T) = H(t)$ is here imposed, with a period T. We introduce quasienergies $\{\omega_l\}$ and quasi-eigenstates $\{|l\rangle\}$ for the unitary operator

$$U = \mathscr{T} \exp\left[-(\mathrm{i}/\hbar) \int_t^{t+T} H(\tau)\, \mathrm{d}\tau \right], \tag{1.53}$$

where $\mathscr{T}$ is a time-ordering operator. The $\{\omega_l\}$ are assumed both nondegenerate and nonresonant. The general solution for (1.52) is then given by

$$\Psi(t = t_0 + NT) = \sum_{l=1}^{\infty} |l\rangle\langle l|\Psi_0\rangle \exp\{-\mathrm{i}NT\omega_l/\hbar\}. \tag{1.54}$$

Using the expression (1.54), the difference d between $\Psi(t)$ and $\Psi(t_0)$ is given by

$$d^2 \equiv |\Psi(t) - \Psi(t_0)|^2 = \textstyle\sum_{\mathrm{I}} + \sum_{\mathrm{II}} \tag{1.55}$$

with

$$\textstyle\sum_{\mathrm{I}} \equiv \displaystyle\sum_{l=1}^{n} 2|\langle l|\Psi_0\rangle|^2\{1 - \cos(NT\omega_l/\hbar)\} \tag{1.56a}$$

and

$$\textstyle\sum_{\mathrm{II}} \equiv \displaystyle\sum_{l=n+1}^{\infty} 2|\langle l|\Psi_0\rangle|^2\{1 - \cos(NT\omega_l/\hbar)\}. \tag{1.56b}$$

Given any small $\varepsilon(>0)$, we can find n satisfying $\sum_{\text{II}} < \varepsilon/2$ because $\sum_{l=1}^{\infty} |\langle l|\Psi_0\rangle|^2 = \langle\Psi_0|\Psi_0\rangle = 1$. Further, noting the discrete nature of $\{\omega_l\}$, we find a dense set $\{N\}$ satisfying $\sum_{\text{I}} < \varepsilon/2$. Consequently, for a dense set of times $\{NT\}$ we have $d^2 < \varepsilon$. In the same way, $|E(t) - E(t_0)| < \varepsilon$ can be shown. Thus the theorem of quantum recurrence has been proved. This theorem applies also to systems with time-independent Hamiltonians (Bocchieri and Loinger, 1957) and for any bounded many-body system. The phenomena of quantum recurrence constitute a genesis of the suppression of quantum chaos (ergodicity and mixing).

The arguments so far still fail to distinguish between quantum systems with (classically) chaotic and regular motions. Further, they have not elucidated the small but finite effects of the Planck constant on the dynamics. We shall find that the initial stage of time evolution can exhibit a clear signature of chaos and that, in the semiclassical limit, some interesting phenomena present at neither the classical nor the quantum limit occur. To see this point, let us examine more carefully the wavefunction features at the initial stage of quantum dynamics. We suppose systems of N degrees of freedom and choose to describe the wavefunctions in terms of minimum-uncertainty states, i.e., coherent states $|\mathbf{p}, \mathbf{q}\rangle$ (Klauder and Skagerstam, 1985). The probability density function is given by

$$P(\mathbf{p}, \mathbf{q}) = (2\pi\hbar)^{-N} |\langle \mathbf{p}, \mathbf{q}|\Psi\rangle|^2, \tag{1.57}$$

which is a quantum analog of the classical distribution function in phase space. The problem of representation is, of course, crucial. In the so-called Wigner representation (Wigner, 1932), the inverse Weyl transformation (Weyl, 1927) is employed and the resultant Wigner distribution function is given by

$$P_{\text{W}}(\mathbf{p}, \mathbf{q}) = (2\pi\hbar)^{-N} \int d\zeta \langle \mathbf{q} - \zeta/2|\Psi\rangle\langle\Psi|\mathbf{q} + \zeta/2\rangle \exp(i\mathbf{p}\cdot\zeta/\hbar). \tag{1.58}$$

Despite its historical significance, $P_{\text{W}}(\mathbf{p}, \mathbf{q})$ can take negative values and show violent undulation of the order of the Planck constant h. In the semiclassical limit, therefore, $P_{\text{W}}(\mathbf{p}, \mathbf{q})$ can neither mimic the classical distribution function (except for a few linear systems such as harmonic oscillators and free particles) nor satisfy the Liouville equation even approximately. All these deficiencies can be overcome by suitable coarse grainings. In fact, coarse grainings imposed by Heisenberg's uncertainty principle are quite reasonable. By a Gaussian smoothing (Husimi, 1940; Weissman and Jortner, 1982; Takahashi and Saito, 1985) of $P_{\text{W}}(\mathbf{p}, \mathbf{q})$, we obtain $P(\mathbf{p}, \mathbf{q})$ in (1.57).

As an initial profile we choose a Gaussian wave packet. Up to the time $\sim\hbar^{-1}$, $P(\mathbf{p}, \mathbf{q})$ in (1.57) mimics well the coarse-grained classical distribution and obeys the Liouville equation. In (classically) nonintegrable and chaotic cases, $P(\mathbf{p}, \mathbf{q})$ begins to develop a stretching and folding (i.e., Smale's horse-shoe) mechanism, changing the wave packet into finer and finer textures.

The growth of complicated patterns can be quantified by using the Kolmogorov–Sinai entropy h_{KS}, by making a correspondence with classical dynamics. To do this, define contour lines $C(t)$ and phase areas $A(t)$ enclosed by them such that the integrated probability $\int_{A(t)} P(\mathbf{p}, \mathbf{q})\, d\mathbf{p}\, d\mathbf{q}$ takes a fixed value (arbitrary). The areas $A(t)$ constitute the incompressible phase liquid, in which each point executes its classical motion. In accord with the wave packet dynamics above, the pattern of $A(t)$ changes from a single spherical droplet to finer and finer maze-like structures (fig. 1.5). If we want to coarsen the fine structures, the phase volume $\Gamma(t)$ covering the overall structures is given by

$$\Gamma(t) = \Gamma_0 \exp(h_{KS} t). \tag{1.59}$$

This indicates that one linear dimension of the phase liquid is maximally extended as $l^{\mathrm{L}} = l_0^{\mathrm{L}} \exp(h_{\mathrm{KS}} t)$. This in turn means that another one orthogonal to l^{L} is maximally contracted as

$$l^{\mathrm{T}} = l_0^{\mathrm{T}} \exp(-h_{\mathrm{KS}} t), \tag{1.60}$$

Fig. 1.5 Time evolution of wave packet.

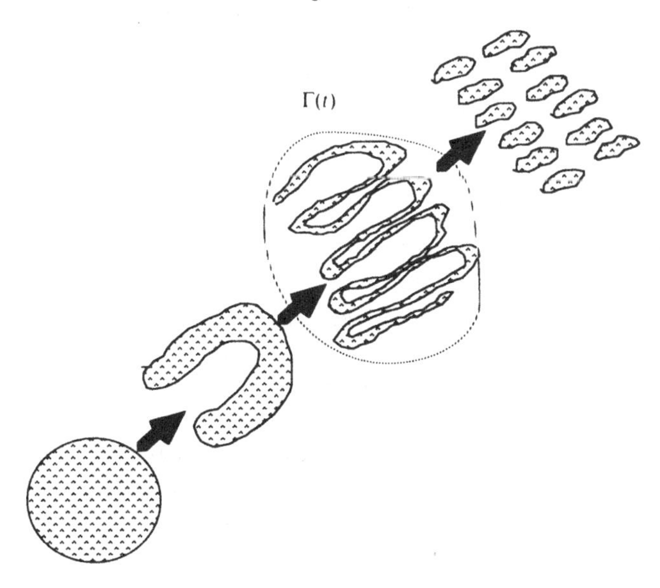

due to Liouville's theorem for the phase liquid. Note that Γ_0, l_0^{L}, $l_0^{\mathrm{T}} = O(1)$.

Because of the uncertainty principle, however, there exists a quantum-mechanical limitation in the resolution of phase space, whose linear dimension is of $O(\hbar^{1/2})$. When l^{T} becomes as small as $\hbar^{1/2}$ at cross-over time τ_c given by

$$\tau_c \simeq h_{\mathrm{KS}}^{-1} \ln(\hbar^{-1}), \tag{1.61}$$

therefore, the classical–quantum correspondence breaks down. For $t > \tau_c$, the dynamics will be dominated by quantum interference, demonstrating a diffusion behavior quite different from that for $t < \tau_c$. Note that the derivation of τ_c in (1.61) is justified since $\tau_c \ll \hbar^{-1}$.

In the semiclassical limit $h \to 0$, in particular, we can observe in the wavefunction the following features: (1) a quantum analog of the stretching and folding mechanism continues to operate for an extended time up to τ_c; (2) beyond τ_c, new morphological patterns on extremely fine scales appear, which may be quantified in terms of fractals or multifractals (see chapter 3).

Thus, even within the framework of the conventional time-dependent Schrödinger equation, quantum dynamics yields a variety of interesting phenomena induced by chaos. Other essential aspects of the quantum dynamics of chaos will be described in chapter 3.

1.6 Chaos versus extrinsic disorder: application to solid-state physics

Our materialistic understanding of natural science leads us to ask the following: in what real systems can we observe the novel aspect of quantum chaos? Before giving a decisive answer to this question, we shall sketch the role of extrinsic (spatial) disorder (e.g., impurities, defects, random potentials and surface roughness) in solid-state physics. Among others, the phenomena of disorder-induced electron localization are the most notable. The importance of the delocalization $\leftrightarrow$ localization transition is comparable to that of the liquid $\leftrightarrow$ glass transition. While the latter is a classical phenomenon for molecular or atomic aggregates, the former is a typical quantum-mechanical effect responsible for quantum transport (e.g., diffusion and electric conductance). Abrahams *et al.* (1979) developed a scaling theory of localization, from which they derived a startling result: for $d \leq 2$, all electron states are localized even by very weak disorder. In the presence of a strong magnetic field, localization is implicated as a possible origin of the integral quantum Hall effect in interface layers in GaAs–AlGaAs heterojunctions and Si MOS (metal–oxide–semiconductor) inversion layers.

Random forces induced by thermal noise are of no less fundamental importance in solid-state physics. This extrinsic (time) disorder constitutes a background determinant of the electric resistance, current and other properties, and, in the problem of spin resonance, gives rise to the nonvanishing relaxation rate, width of absorption lines, etc.

Nowadays, however, we can fabricate conducting devices of sufficiently high purity to suppress the spatial disorder. Also, progress in low-temperature technology has enabled us to suppress the effect of thermal noise. Therefore, we are facing a new era in which we will be able to see phenomena caused purely by deterministic randomness or chaos at the microscopic level. In mesoscopic conducting devices with electronic mean free path larger than the sample size (about 1 μm), for instance, electrons show ballistic (rather than diffusive) motion and the electrical conductance is mainly controlled by boundary effects, i.e., billiard boundaries. So we shall see the effect of quantum chaos on quantum transport. Another example of chaos in the microscopic world can be found in magnetic resonance in ferrites, where the origin of the anomalous broadening of absorption lines can be attributed to chaos rather than to thermal noise (see chapter 4).

The analogy between chaos and extrinsic randomness was first illustrated in the context of the kicked rotator (Casati *et al.*, 1979; Fishman *et al.*, 1982), which describes an electron subjected to a periodically pulsed (electric) field. Its Hamiltonian is given by

$$H = p^2/2 + k \cos(q) \sum_{n=-\infty}^{\infty} \delta(t - nT), \tag{1.62}$$

where k and T denote the strength of the pulse and the period of pulses, respectively. The corresponding classical dynamics is captured by the two-dimensional standard map (Chirikov, 1979) for canonical variables $\{p_n, q_n\}$ just before the successive pulses:

$$p_{n+1} = p_n + k \sin(q_n), \qquad q_{n+1} = q_n + T \cdot p_{n+1}. \tag{1.63}$$

By a suitable rescaling, the map actually depends on the parameter $K(\equiv kT)$. The threshold (Greene, 1979) at which the last KAM torus collapses and the transition to global chaos occurs is $K_{\text{th}} = 0.9716\cdots$. For $K \gg K_{\text{th}}$ in the chaotic regime, the diffusion coefficient is calculated as

$$D = \lim_{N\to\infty} \langle (p_N - p_0)^2 \rangle/(NT) = T^{-1} \lim_{N\to\infty} N^{-1} \left\langle \left[k \sum_{n=0}^{N-1} \sin(q_n) \right]^2 \right\rangle$$
$$= k^2/(2T), \tag{1.64}$$

where $\langle \cdots \rangle$ denotes the average over initial p_0 values.

As noted in section 1.5, the quantum-mechanical treatment of chaos is destined to lead to the suppression of chaotic diffusion. In the case of the quantum kicked rotator, this phenomenon can also be explained in the context of localization in solid-state physics. The problem of quantum dynamics for (1.62) is simply reduced to solving the eigenvalue problem for the one-period propagator

$$U|\Psi\rangle \equiv \mathrm{e}^{-(\mathrm{i}k/\hbar)\cos(q)}\cdot \mathrm{e}^{-(\mathrm{i}/2\hbar)Tp^2}|\Psi\rangle = \mathrm{e}^{-(\mathrm{i}/\hbar)T\omega}|\Psi\rangle, \tag{1.65}$$

where ω is a quasi-eigenvalue (see (1.53) and (1.54)). There are several ways to tackle (1.65). Here, we introduce a suitable multiplication factor $g(q)$ and take an expansion of $|\Psi\rangle$ in momentum space as

$$g(q)|\Psi\rangle = \sum_{r=-\infty}^{\infty} c_r\,\mathrm{e}^{\mathrm{i}rq}. \tag{1.66}$$

For a choice $g(q) = 2^{-1}\{1 + \exp[(\mathrm{i}k/\hbar)\cos(q)]\}$, (1.65) is reduced to

$$T_m c_m + \sum_n W_n c_{m+n} = 0, \tag{1.67}$$

where $T_m = \tan[\gamma\pi m^2 - (\omega T/2\hbar)]$ with rationality $\gamma \equiv \hbar T/(4\pi)$ and $W_n = (1/2\pi)\int_0^{2\pi} \mathrm{d}q\, \mathrm{e}^{\mathrm{i}nq}\tan[(k/2\hbar)\cos q]$. Equation (1.67) is analogous to a one-dimensional tight-binding model for an electron in a crystalline solid, if m and n are taken as site indices. In this context, T_m and W_n are taken as site energy and hopping matrix elements, respectively. In the case of resonance (i.e., γ = rational), T_m is periodic in m and all eigenstates of (1.67) are extended Bloch states. In more general cases of nonresonance (i.e., γ = irrational), $\{T_m\}$ is pseudo-random. Equation (1.67) now resembles a one-dimensional Anderson model for localization with all the states showing an exponential decay in the momentum space as

$$|c_n|^2 \sim \exp\{-(2/\xi)|n - n_0|\}, \tag{1.68}$$

where ξ is the localization length. ξ is now related to the diffusion constant (1.64) by

$$\xi = [\hbar T/(2\pi)]^2 D. \tag{1.69}$$

The energy spectrum here is singular-continuous. Numerical evidence for (1.68) and (1.69) is given in fig. 1.6. The quantum analog of chaotic diffusion ceases at cross-over time τ_c in (1.61). Therefore this localization allows us to understand why the quantum suppression of chaos occurs.

To speak rigorously, the correspondence with the localization model in solid-state physics will break down for $(k/2\hbar)\cos q \geq \pi/2$, because W_n in (1.67) becomes divergent. Shepelyansky (1986) has removed this difficulty, by choosing another expression for $g(q)$ in (1.66) as

$g(q) = \exp[-(ik/2\hbar)\cos q]$: the resultant eigenvalue problem is reduced to

$$\sum_{n=-\infty}^{\infty} J_n(k/2) \sin\left(\gamma\pi m^2 + \frac{n\pi}{2} - \frac{\omega T}{2\hbar}\right) c_{m+n} = 0, \tag{1.70}$$

where J_n is the nth order integral Bessel function. Equation (1.70) is free from the problem of divergence and can be related to the Lloyd model, which is another typical model of localization.

In this way we can recognize a great similarity between the quantum kicked rotator and the one-dimensional tight-binding model for localization in solid-state physics. It should however be noted that the kicked rotator is not a typical model of quantum chaos. In general (classically) chaotic systems, most of the wavefunctions are rather delocalized in an ergodic way. Though periodic orbit scars (Heller, 1984) can be perceived in some systems (e.g., the stadium billiard system), they are embedded in the ergodic sea of wavefunctions and do not constitute localized states. Further, localization of the kicked rotator in momentum space has nothing to do with the actual Anderson localization in real space. We shall instead suggest more direct evidence for quantum chaos in solid-state physics.

Anderson localization of electrons is indeed recognized to occur in the presence of random potentials, but the meaning of 'random' potentials should be scrutinized. Nothing can legitimately be random unless it is the

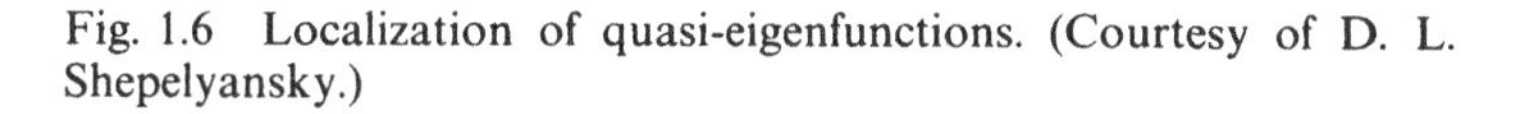

Fig. 1.6 Localization of quasi-eigenfunctions. (Courtesy of D. L. Shepelyansky.)

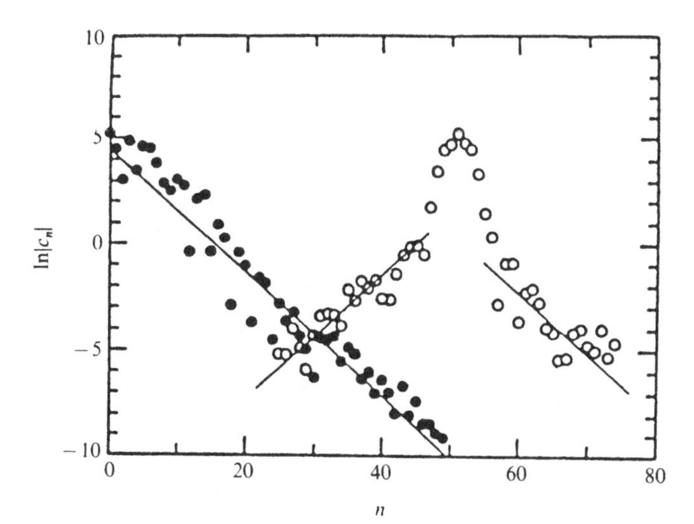

member of an ensemble. The potential in a dirty metal is unique and not random at all. There is no way of constructing an ensemble. This dilemma will be resolved by introducing the concept of quantum chaos. Laughlin (1987) suggested a novel idea on the physics of quantum transport in a *diffusive regime*. While finite electrical conductivity has traditionally been understood as caused by the motion of electrons in random potentials in dirty metals, Laughlin guesses it to be induced by deterministic chaos. Consider a crystalline array of hard spheres (radius R) with unit-cell volume Ω, i.e., a three-dimensional version of the Sinai billiard (see fig. 1.7). The classical motion of an electron in this system is fully chaotic. For constant electron velocity v, the Lyapunov exponent λ can be estimated as follows. The rate at which the electron collides with spheres is

$$\tau^{-1} = v\pi R^2/\Omega. \tag{1.71}$$

An uncertainty Δx in the impact parameter yields, by a collision, the velocity uncertainty in the x direction as

$$\Delta v \sim v\,\Delta x/R. \tag{1.72}$$

During time τ, Δx grows to

$$\Delta x' = \Delta x + \Delta v\,\tau \sim (1 + v\tau/R)\,\Delta x. \tag{1.73}$$

Combining (1.73) with $\Delta x'/\Delta x \sim \exp(\lambda\tau)$ and noting (1.71) and $R^3 \sim \Omega$, we find

$$\lambda \sim \tau^{-1}. \tag{1.74}$$

Fig. 1.7 Motion of classical electron in Sinai billiard.

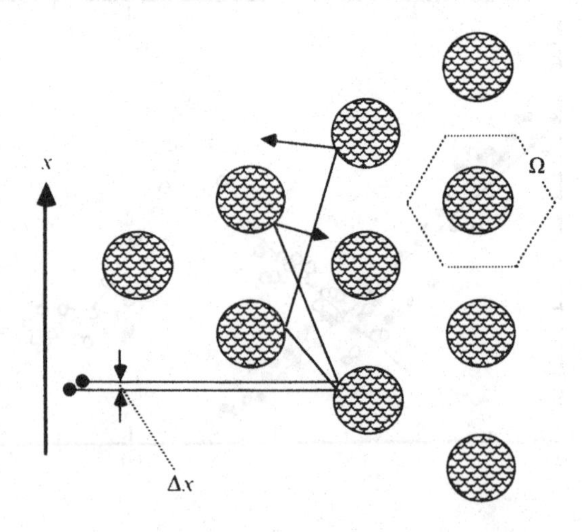

Equation (1.74) implies that chaos renders the electron motion diffusive. In the quantum-mechanical treatment, the Lyapunov exponent is meaningful only conditionally (see section 1.5), but let the experiment be performed within a period shorter than the natural time scale associated with the density of states $\hbar\mathscr{D}(\varepsilon)$. Then the quantum (position) uncertainty after one collision can be less than the distance between scatterers and a clear definition of the Lyapunov exponent is still possible.

The rate of decay of velocity correlations may be identified with λ. The electrical conductivity of a gas of classical electrons is then given by

$$\sigma(\omega) = \frac{\rho e^2}{m} \int_0^\infty \frac{\langle v(0) v(t) \rangle}{\langle v(0) v(0) \rangle} \mathrm{e}^{\mathrm{i}\omega t}\, \mathrm{d}t, \tag{1.75}$$

where ρ, e and m are the electron density, charge and mass, respectively. Equation (1.75) corresponds to the Kubo formula for the conductivity of a degenerate noninteracting electron gas. Using $\langle v(0)v(t)\rangle = \langle v(0)v(0)\rangle\, \mathrm{e}^{-\lambda t}$ in (1.75), we obtain the Drude conductivity in terms of the Lyapunov exponent:

$$\sigma(\omega) = \frac{\rho e^2}{\lambda m} \frac{1}{1 - \mathrm{i}\omega/\lambda}, \tag{1.76}$$

which is valid for $\omega > [\hbar\mathscr{D}]^{-1}$, i.e., for short-time experiments. Though real potentials in metals cannot be exactly simulated by the Sinai billiard, the above scenario indicates an intriguing role for deterministic chaos in quantum transport.

Quantum chaos thus provides a possible mechanism for the random motion of electrons. While, in the scaling theory of localization (Abrahams *et al.*, 1979), the conductivity is a fundamental measure of the randomness of the potential, it can now be seen as a fundamental measure of the chaotic instability of electronic motions. As we shall see in chapter 2, the quantum transport in a *ballistic regime* is much more controlled by quantum chaos. Realization of quantum chaos in more diverse fields of solid-state physics is possible, e.g., in mesoscopic electronic devices and in magnetic resonance. Detailed description of these subjects will be given in following chapters.

2
Quantum billiards: from closed to open systems

In chapters 2–4, we provide several realistic examples of quantum chaos in condensed-matter physics. In this chapter, we are concerned with autonomous systems involving an orbital degree of freedom for electrons. The classical and quantum mechanics of noninteracting electrons are studied in a nonintegrable elliptic billiard with a uniform magnetic field applied perpendicularly. We attempt to establish a connection between chaotic dynamics and macroscopic quantum observables. First, after an overview of the long history of orbital diamagnetism (from Bohr through Landau and Peierls), we point out the significance of Van Leeuwen's thought for modern nonlinear dynamics. We see how the classical skipping motion of electrons becomes chaotic, and then we proceed with a search for the quantum-mechanical counterpart of chaos. For small two-dimensional systems, a remarkable reduction in, and large fluctuations of, the diamagnetic susceptibility are found in the corresponding classically chaotic regime. In particular, the close relationship between avoided level crossings and diamagnetism is asserted. In the final section, on suppressing the magnetic field, we embark on the very topical study of open stadium billiards with a pair of conducting lead wires, examining the novel effects of transient chaos on quantum transport phenomena such as electrical conductance.

2.1 Diamagnetism of a Fermi gas: revival of Van Leeuwen's thought

Orbital diamagnetism of electrons is one of the key concepts for understanding some prominent phenomena in solid-state physics. For instance, superconductivity is indicated by complete diamagnetism (the Meissner effect) and the integral quantum Hall effect is known to be closely related to diamagnetic edge currents. The history of the theory of orbital diamagnetism (Mott and Jones, 1936; Ashcroft and Mermin, 1976) can be traced back to Niels Bohr's Ph.D. dissertation (1911), in which he tried to explain this phenomenon from a classical viewpoint. Van Leeuwen later developed a more comprehensive analysis (1921). These monumental works are worthy of mention here in some detail, even though both scholars failed to provide an adequate explanation of orbital diamagnetism, owing to limitations in classical mechanics.

In the classical treatment, an electron (with mass m and charge $-e$) in the presence of a magnetic field $\mathbf{B}$ obeys

$$m\,\mathrm{d}^2\mathbf{r}/\mathrm{d}t^2 = -(e/c)\mathbf{v} \times \mathbf{B}, \tag{2.1}$$

and executes a helical motion along the direction of $\mathbf{B}$, with Larmor radius $R = mvc/eB$ and frequency $\omega_c = eB/mc$. This orbital motion in turn brings about a diamagnetic moment $\mu = (eR^2/2c)\omega_c$ anti-parallel to $\mathbf{B}$. Nevertheless, the classical-statistical mechanics for the electron gas as a whole always leads to a prediction of vanishing diamagnetic susceptibility! The proof is quite simple. The Hamiltonian for the electron gas in a magnetic field $\mathbf{B} = \nabla \times \mathbf{A}$ is

$$\mathscr{H} = \sum \frac{1}{2m}\left\{\mathbf{p}_j + \frac{e}{c}\mathbf{A}(\mathbf{r}_j)\right\}^2 + U(\mathbf{r}_1, \mathbf{r}_2, \ldots, \mathbf{r}_N). \tag{2.2}$$

The corresponding partition function is given by

$$Z = \int \mathrm{d}^3\mathbf{r} \int \mathrm{d}^3\mathbf{p} \exp\left(-\frac{\mathscr{H}}{kT}\right). \tag{2.3}$$

By making the variable change $\mathbf{p} \to \boldsymbol{\pi} = \mathbf{p} + (e/c)\mathbf{A}$, the volume element in (2.3) becomes $\mathrm{d}^3\mathbf{r}\,\mathrm{d}^3\boldsymbol{\pi}$. Therefore, Z and the resultant free energy F become thoroughly $\mathbf{A}$-independent and thereby $\mathbf{B}$-independent. So the susceptibility is vanishing ($\chi = -\partial^2 F/\partial B^2 = 0$), which cannot be reconciled with the experimental facts. In an attempt to overcome the puzzle, Van Leeuwen left us an interesting proposal: gyrating electrons collide with container walls (metal), yielding the edge current which cancels the bulk current arising from complete Larmor orbits (see fig. 2.1). Van Leeuwen's proposal is indeed suggestive and has acquired renewed interest quite recently in experimental studies of mesoscopic conducting disks and in the context of nonlinear dynamics and chaos in magnetic billiards. This point will be examined systematically in later sections.

Soon after the establishment of quantum mechanics, Landau (1930) provided the first theory compatible with experiments, treating the three-dimensional system at finite temperature. For the sake of simplicity, however, we concentrate here on a two-dimensional electron gas at absolute zero and describe Peierls' intuitive procedure (1933a, 1933b). (Effects of spin degree of freedom and lattice discreteness are also suppressed.)

Quantized energies (Landau, 1930) are given by

$$E_n = (1/2)\hbar\omega_c(2n+1) \qquad (n = 0, 1, \ldots). \tag{2.4}$$

The Landau levels E_n are degenerate with respect to the center of Larmor's circular orbits. The degree of degeneracy or statistical weight of each level is commonly γB with an appropriate constant γ, since it is given by the sample area divided by the area of the Larmor orbit $2\pi R_0^2$. (R_0 is the radius of the lowest Landau state and is given by $R_0^2 = \hbar/eB$.)

Let the Fermi energy ε_F lie in the lth Landau level E_l. The ground-state energy E_G is then given by

$$\begin{aligned} E_G &= \gamma B \sum_{n=0}^{l-1} (\hbar\omega_c/2)(2n+1) + (\hbar\omega_c/2)(N_F - l\gamma B)(2l+1) \\ &= (\hbar eB/2mc)\{(2l+1)N_F - \gamma Bl(l+1)\}, \end{aligned} \tag{2.5}$$

where N_F is the number of electrons filling up to ε_F levels. From (2.4), we have the susceptibility χ at absolute zero

$$-\chi = \partial^2 E_G/\partial B^2 = (e\hbar/mc)\gamma l(l+1). \tag{2.6}$$

Noting that $l = [N_F/\gamma B]$ (Gauss' notation) owing to the inequality $l\gamma B < N_F < (l+1)\gamma B$, $-\chi$ is found to be proportional to B^{-2} in the low-field

Fig. 2.1 Van Leeuwen's bulk and edge currents.

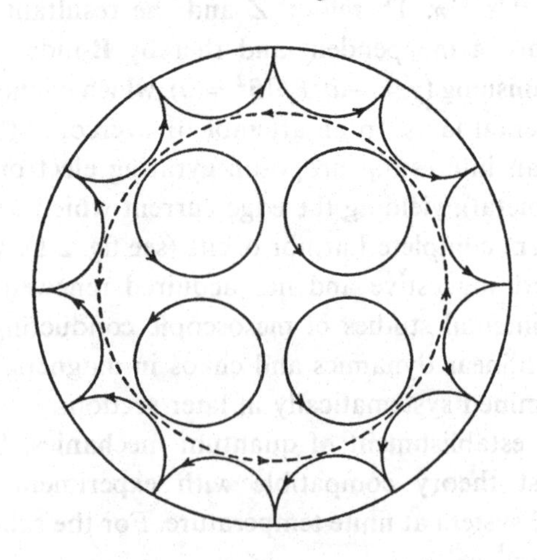

regime and to B^{-1} in the strong-field regime (see fig. 2.2) – a typical feature of a two-dimensional version of Landau diamagnetism. While more elaborate studies on Landau diamagnetism have concerned effects of lattice discreteness, Zeeman splitting due to spin degree of freedom, etc., little attention has been paid to the effect of hard walls at the boundary. In fact, most existing studies are based on Landau levels, broadened or not; but Landau levels are available only by suppressing boundary effects. If Van Leeuwen's assertion is taken into consideration in the quantum mechanical treatment, then Landau levels are no longer meaningful and we shall face unexpected issues.

2.2 Classical dynamics in magnetic billiards

The study of boundary effects on the dynamics of a charged particle in a magnetic field is nothing but dealing with magnetic billiards. This study began quite recently in the context of nonlinear dynamics and chaos. Nakamura and Thomas (1988) developed a quantum-mechanical study. Before describing this study, we introduce the results of the classical-mechanical treatment by Robnik and Berry (1985).

Bouncing orbits can be considered geometrically as a sequence of arcs of Larmor radius $R = mvc/eB$. Without loss of generality, orbits are taken to gyrate clockwise. As seen in fig. 2.3, each bouncing point on the

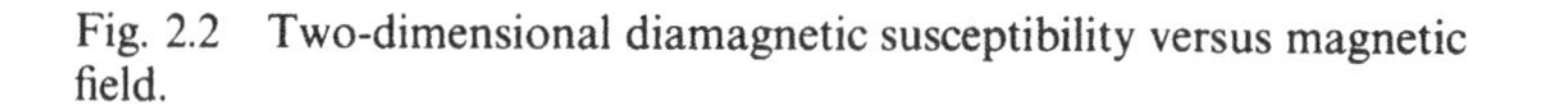

Fig. 2.2 Two-dimensional diamagnetic susceptibility versus magnetic field.

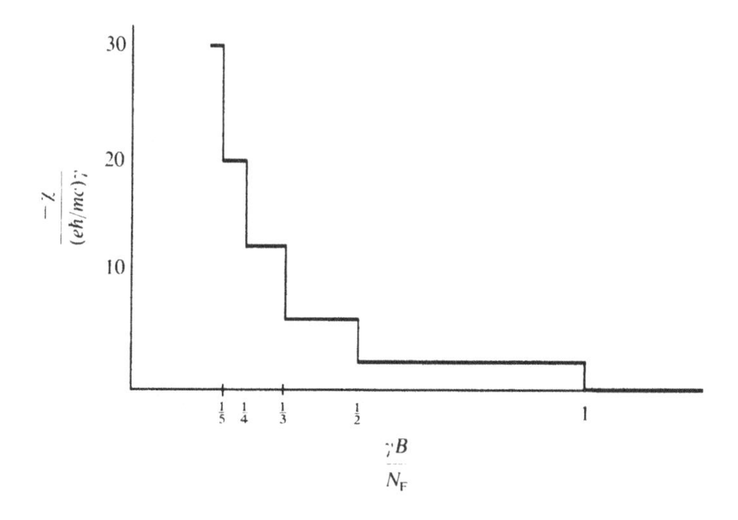

boundary is labeled by arc length s measured along the boundary with perimeter $\mathfrak{L}$; $0 \leq s \leq \mathfrak{L}$. The angle of emergence Θ is measured from the forward tangent to the boundary ($0 \leq \Theta \leq \pi$). With a choice of tangential momentum $p = \cos \Theta$, a two-dimensional bounce map for successive values (s_n, p_n) can be constructed.

Behaviors of orbits depend strongly on R and are classified in terms of three characteristic lengths given in fig. 2.4: R^* for the radius of the largest

Fig. 2.3 Notation for successive bouncings.

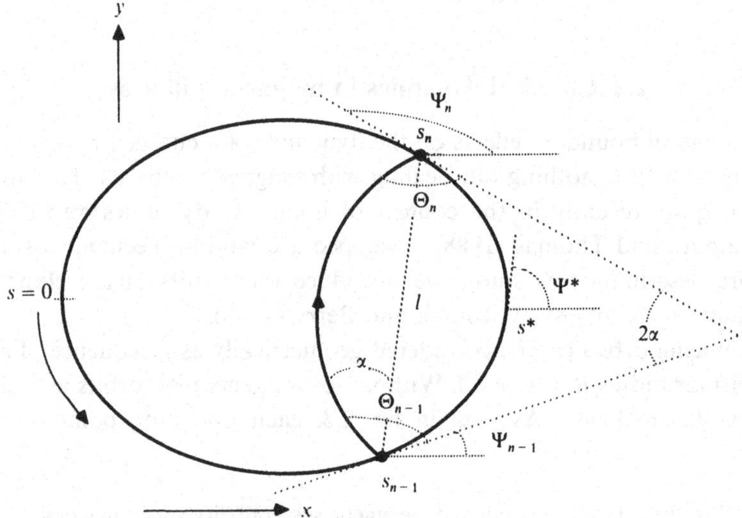

Fig. 2.4 Definition of R^*, ρ_{min} and ρ_{max}.

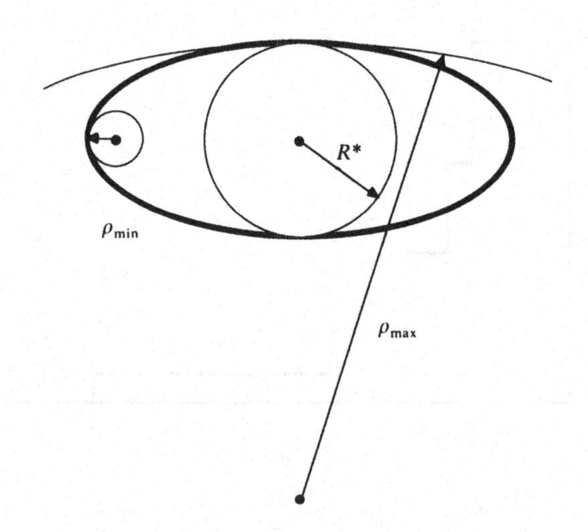

inscribed circle; ρ_{min} and ρ_{max} for the radii of the maximum and minimum curvatures of the boundary, respectively. These lengths satisfy $\rho_{min} < R^* < \rho_{max}$. (1) For $\rho_{max} < R < \infty$, the motion is close to that in the field-free case. (2) For $\rho_{min} < R < \rho_{max}$, the grazing circle intersects the billiard, showing flyaway behavior. (3) For $R < \rho_{min}$, we find orbits skipping along the boundary. Figure 2.5 shows initial stages of typical orbits. In addition to the bouncing orbits above, complete Larmor orbits (closed circles) begin to coexist when R is decreased below R^*.

The Kolmogorov–Arnold–Moser theorem indicates the presence of adiabatic invariants near the integrable regime. When the arc length interval Δs between successive bounces is much less than $\rho(s)$, the local radius of curvature, the curvature $\kappa(s)$ can be taken as changing 'adiabatically' such that the linear approximation

$$\kappa(s) \simeq \kappa_0 + \kappa_1(s - s^*) \tag{2.7}$$

is valid. In this limit, we can reasonably anticipate the presence of the conserved function $I(s, p)$. Using tangent directions Ψ_{n-1}, Ψ_n and Ψ^* and other angles Θ_n, Θ_{n-1}, α shown in fig. 2.3, we find

$$\Psi^* = \Psi_{n-1} + \Theta_{n-1} - \alpha = \Psi_n - \Theta_n + \alpha. \tag{2.8}$$

From (2.8), $\Delta\Theta = \Theta_n - \Theta_{n-1}$ is given by

$$\Delta\Theta = \Psi_n + \Psi_{n-1} - 2\Psi^*. \tag{2.9}$$

Fig. 2.5 Initial stage of typical orbits: (*a*) $R = \infty$; (*b*) $R \gg \rho_{min}$; (*c*) $R < \rho_{min}$.

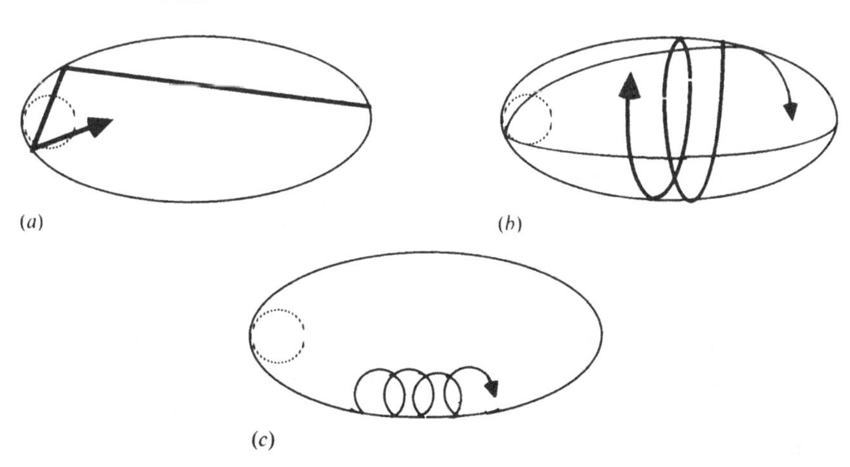

$\Psi(s)$ can be expressed here in terms of κ:

$$\Psi(s) = \Psi^* + \int_{s^*}^{s} ds' \, \kappa(s'). \tag{2.10}$$

In contrast, s_{n-1} and s_n are determined by the condition

$$\int_{s_{n-1}}^{s_n} ds \sin(\Psi^* - \Psi) = 0, \tag{2.11}$$

which implies the vanishing of the integrated perpendicular distance from chord to boundary.

Using (2.7) in (2.10), (2.11) and (2.9), we arrive at

$$\Delta\Theta \simeq \kappa_1(\Delta s)^2/6 = \frac{\Delta\kappa}{\Delta s}(\Delta s)^2/6. \tag{2.12}$$

Δs can be found by approximating the local boundary with a circle of radius $\kappa_0^{-1} \simeq \kappa^{-1}$ as

$$\Delta s \simeq 2R \sin\Theta/(1 + \kappa R \cos\Theta). \tag{2.13}$$

The combination of (2.12) and (2.13) together with a variable change ($\Theta \to p$) results (in the limit $\Delta s \to 0$) in

$$\frac{dp}{d\kappa} = -\sin\Theta \frac{d\Theta}{ds}\left(\frac{d\kappa}{ds}\right)^{-1} = -R(1 - p^2)[3(1 + \kappa R p)]^{-1}, \tag{2.14}$$

integration of which gives rise to the desired adiabatic invariant:

$$I(s, p) = [R\kappa(s) + p(3 - 2p^2)](1 - p^2)^{-3/2}. \tag{2.15}$$

This invariant can survive for very short skips, i.e., $\Delta s \ll 1$ in (2.13). In the weak-field limit $\kappa R \to \infty$, (2.15) reduces to known results associated with caustics close to the boundary (Lazutkin, 1973; Keller and Rubinow, 1960). The above result is valid for any concave billiard with smooth boundary.

To see the breakdown of adiabatic invariants, i.e., collapse of KAM tori and occurrence of chaos, the shape of the billiard is now specified as an ellipse and Newton's equation of motion is solved numerically. For an ellipse with semi-major and semi-minor axes a and b, respectively, we choose a 'control parameter' $\sigma = b/a$ such that $a = \sigma^{-1/2}$ and $b = \sigma^{1/2}$. Then, we have the eccentricity $(1 - b^2/a^2)^{1/2} = (1 - \sigma^2)^{1/2}$ and

$$\rho_{\min} = \sigma^{3/2}, \tag{2.16a}$$
$$\rho_{\max} = \sigma^{-3/2}, \tag{2.16b}$$
$$R^* = \sigma^{1/2}. \tag{2.16c}$$

In the zero-field case $R = \infty$, the system is integrable (Sinai, 1976) and the phase space is covered with invariant tori (see fig. 2.6(*a*)). Two elliptic fixed points $(s, p) = (\mathfrak{L}/4, 0)$, $(3\mathfrak{L}/4, 0)$ and three hyperbolic fixed points $(s, p) = (0, 0)$, $(\mathfrak{L}/2, 0)$, $(\mathfrak{L}, 0)$ correspond to short- and long-diametrical orbits in fig. 2.5(*a*), respectively.

In the nonvanishing-field case, with $R \gg \rho_{\min}$, orbits around the separatrix become chaotic (see figs. 2.5(*b*) and 2.6(*b*)), but the chaotic fraction occupies only part of the phase space, with the remainder around $p = \pm 1$ occupied by invariant curves given by (2.15). For stronger field ($R < R^*$), a new chaotic region appears around $p = -1$ (see figs. 2.5(*c*) and 2.6(*c*)), while the old chaotic region begins to shrink. Finally, for $R \ll \rho_{\min}$, chaos dies out and the phase space is again filled by invariant tori, as in fig. 2.6(*d*).

Fig. 2.6 Poincaré sections based on bounce map. (*a*) $R = \infty$; (*b*) $R = 3$; (*c*) $R = 0.75$; (*d*) $R = 0.4$. (Courtesy of M. Robnik and M. V. Berry.)

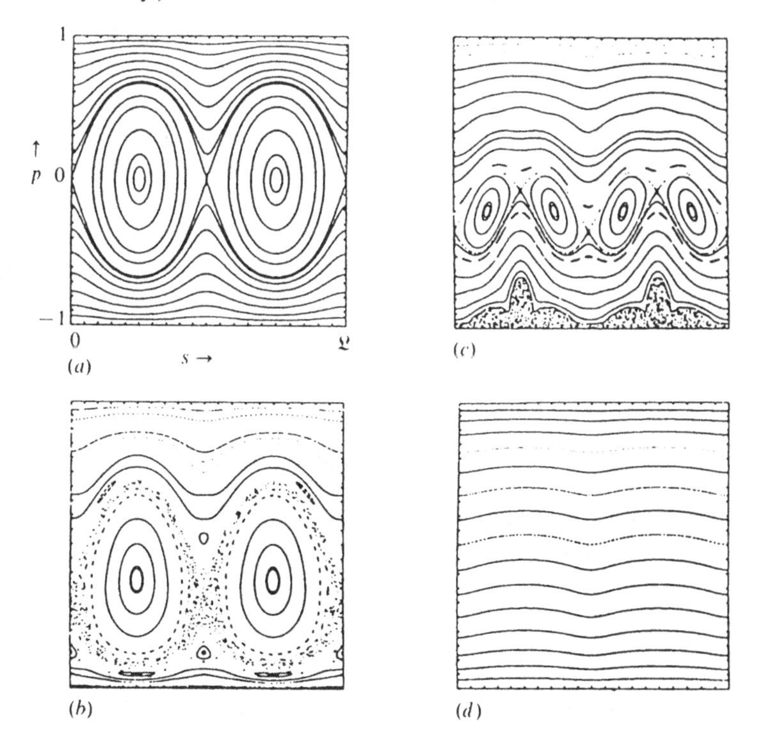

Despite the variety of results above, Robnik and Berry stayed within classical mechanics, performing no analysis of the corresponding quantum mechanics. What kind of novel quantum effects will appear for $R > \rho_{\min}$? How will Landau diamagnetism be violated by incorporating the effect of chaos? The answers can be found in the next section.

2.3 Avoided level crossings and diamagnetism

We now discuss the quantum-mechanical treatment of noninteracting electrons in a planar billiard (e.g., a thin conducting disk) in a uniform magnetic field normal to the plane (Nakamura and Thomas, 1988). The shape of the boundary is taken as elliptic. Unless the boundary effects are taken into consideration, we merely obtain Landau levels and associated Landau diamagnetic susceptibility, as pointed out in section 2.1. Single-electron classical dynamics in section 2.2, however, elucidated the onset of chaos in the case where the cyclotron radius is comparable to the linear dimension of the billiard, indicating the crucial role of the concave boundary.

Quantum aspects of chaos can be captured by the incorporating of Dirichlet-type boundary conditions. In this section, we first solve the Dirichlet eigenvalue problem for a single-electron system. Then, global aspects of the spectra, i.e., the sensitivities of energies and their average over 'occupied' levels – the diamagnetic susceptibility at absolute zero – will be investigated by changing the magnetic field. Effects of lattice discreteness and spin degrees of freedom will be suppressed in the present treatment.

Let us consider an ellipse of area $\pi L^2 = \pi ab$ where a and b are semi-major axes in the x and y directions, respectively. The eigenvalue problem is given by

$$H\Psi = E\Psi \tag{2.17}$$

with $\Psi = 0$ at the boundary. Here

$$H = (1/2m)[(\hbar/\mathrm{i})\nabla + (e/c)\mathbf{A}]^2. \tag{2.18}$$

We take a symmetric gauge: $\mathbf{A} = (-\frac{1}{2}yB, \frac{1}{2}xB)$ with B the magnetic field. The present system has C_2 (inversion) symmetry. Let us consider the map $(x, y,) \to (r, \theta)$ via $x = ar\cos\theta$, $y = br\sin\theta$. Then, the eigenvalue problem is reduced to

$$\tilde{H}(r, \theta)\tilde{\Psi}(r, \theta) = E\tilde{\Psi}(r, \theta) \tag{2.19}$$

with $\tilde{\Psi} = 0$ at the boundary of the unit disk. Basis functions are now constructed in terms of integer-order Bessel functions as

$$\{|kn\rangle\} \equiv \{R_{kn} J_k(\gamma_{kn} r)\, e^{ik\theta}\} \tag{2.20}$$

where γ_{kn} are zeros of $J_k(z)$ and

$$R_{kn} \equiv [\pi^{1/2} J_{k+1}(\gamma_{kn})]^{-1} \tag{2.21}$$

are normalization constants. The $\{|kn\rangle\}$ are arranged in order of increasing values of γ_{kn}^2. Nonvanishing matrix elements of $\tilde{H}$ are given in dimensionless form as

$$(\hbar^2/2mL^2)^{-1}\langle k'n'|\tilde{H}|kn\rangle = (\pi/2)(1+\sigma^2)\{[2\gamma_{kn}^2/(\bar{a}\sigma)^2 + 2\bar{B}k/\sigma]\Gamma^{(1)}_{kn'kn} + (\bar{B}^2\bar{a}^2/2)\Gamma^{(3)}_{kn'kn}\} \tag{2.22a}$$

for $k' = k$ and

$$(\pi/2)(1-\sigma^{-2})\{-2[(k+1)/\bar{a}^2]\gamma_{kn}\Lambda^{(0)}_{k'n'kn} + (\gamma_{kn}^2/\bar{a}^2)\Gamma^{(1)}_{k'n'kn} + (\bar{B}\gamma_{kn}/\sigma)\Lambda^{(2)}_{k'n'kn} + (\bar{B}^2\bar{a}^2/4)\Gamma^{(3)}_{k'n'kn}\} \tag{2.22b}$$

for $k' = k + 2$ together with their real symmetric counterparts ($k' \leftrightarrow k$, $n' \leftrightarrow n$). Here

$$\Gamma^{(l)}_{k'n'kn} \equiv R_{k'n'} R_{kn} \int_0^1 r^l J_{k'}(\gamma_{k'n'} r) J_k(\gamma_{kn} r)\, dr \tag{2.23}$$

and $\Lambda^{(l)}_{k'n'kn}$ is given by the replacement of the last factor in this integrand by $J_{k+1}(\gamma_{kn} r)$. We have also introduced the dimensionless parameters $\sigma = b/a$, $\bar{a} = a/L$, $\bar{b} = b/L$ and $\bar{B} = B(c\hbar/eL^2)^{-1}$.

From these matrix elements, we recognize the following: (i) the Hilbert space is decomposed into two subspaces of even ($k = 0, \pm 2, \ldots$) and odd ($k = \pm 1, \pm 3, \ldots$) parity; (ii) in the case $\sigma = 1$, k is a good quantum number (i.e., angular momentum); (iii) for $\bar{B} = 0$ with $\sigma = 1$, eigenvalues reduce to

$$E = (\hbar^2/2mL^2)\gamma_{kn}^2. \tag{2.24}$$

In each of the two subspaces, we have taken the lowest 150 basis functions and computed the integrals $\Gamma^{(l)}_{k'n'kn}$ and thereby $\langle k'n'|\tilde{H}|kn\rangle$. Then, we have solved the eigenvalue problem for the dimensionless matrix above, separately in each manifold, to obtain scaled energies $\bar{E} = E(\hbar^2/2mL^2)^{-1}$, in the range $0 < \sigma \leq 1$ and $0 \leq \bar{B} \leq 50$. While changing σ, the area of the billiard has been kept constant: $\bar{a} = 1/\sigma^{1/2}$, $\bar{b} = \sigma^{1/2}$. We have checked the reliability of decimal places of the eigenvalues by comparing them with the corresponding values obtained from enlarged (200×200) matrices.

We find that use of about the lowest 50 eigenvalues for each parity gives sufficient precision for our study. We present below the results for a nonintegrable case, $\sigma = 0.5$ (elliptic billiard), and the integrable case $\sigma = 1$ (circular billiard). Comparison of the two cases helps to elucidate the effects of nonintegrability.

In fig. 2.7, the even-parity part of the energy spectra is shown. In both figs. 2.7(*a*) and 2.7(*b*), most of the levels are found not to be well bunched into Landau levels. Level repulsion leading to avoided crossings is widely seen in the case $\sigma = 0.5$, see fig. 2.7(*b*), while true crossings between levels with different k values predominate in the case $\sigma = 1$, see fig. 2.7(*a*).

The presence of many avoided crossings corresponds to chaos in the underlying classical dynamics. The latter was analyzed in section 2.2 by the two-dimensional bounce map for values of the arc length and skipping angle at successive bounces of an electron at the boundary. For $\sigma \neq 1$, chaos around the orbit of unstable diameter and/or flyaway chaos are found to dominate phase space, provided that the Larmor radius R satisfies

$$R/L \geq \rho_{\min} \tag{2.25}$$

for the smallest curvature radius

$$\rho_{\min} = \bar{b}^2/\bar{a}. \tag{2.26}$$

Fig. 2.7 $\bar{B}$-dependent energy spectra for the even-parity manifold: (*a*) circle billiard ($\sigma = 1$); (*b*) ellipse billiard ($\sigma = 0.5$).

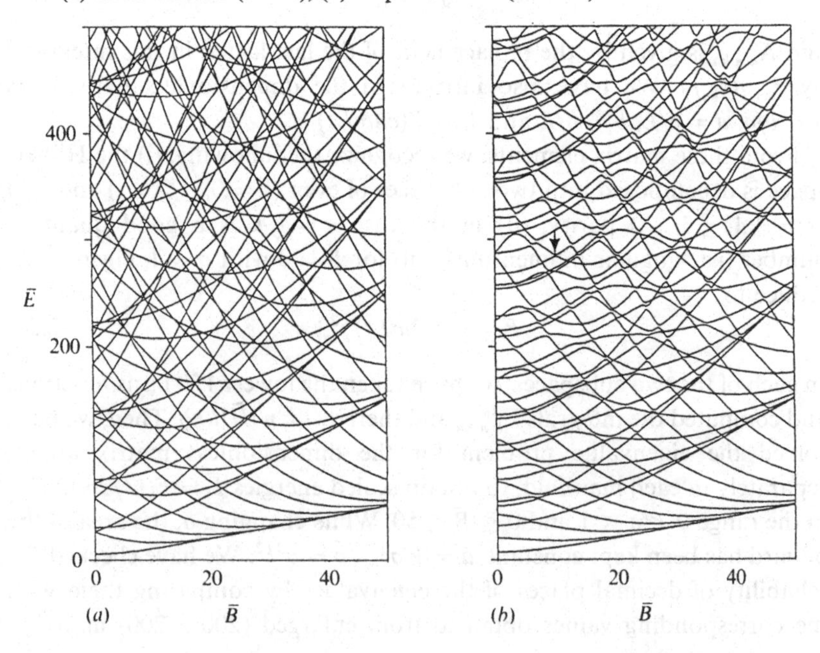

Noting that $E = \frac{1}{2}mv^2$ for electron velocity v, we find

$$R/L = \bar{E}/\bar{B}^2. \tag{2.27}$$

Consistent with the classical findings, we clearly observe in fig. 2.7(*b*) that avoided crossings dominate spectra in the region $\bar{E}/\bar{B}^2 \geq 0.35$ ($= \rho_{\min}$) for $\sigma = 0.5$. More careful examination indicates that most of the avoided crossings in this region have much broader width than those discernible in the opposite region.

We consider a sample containing $2N$ electrons, and neglect the Zeeman splitting of the spin states. Then, in the free-electron ground state at a given value of the applied field, the lowest N levels $E_j(B)$ ($j = 1, \ldots, N$) are filled with two electrons each, the Fermi levels ε_F lying between $E_N(B)$ and $E_{N+1}(B)$. The isothermal susceptibility per electron at absolute zero is given by the second-order derivative (difference in our computations) of the total energy as (Peierls, 1933a, 1933b, 1979)

$$\chi = -(2N)^{-1}\Delta^2\left[2\sum_{j=1}^{N} E_j\right]\Big/\Delta B^2 = -\frac{2mL^2}{\hbar^2}\mu_B^2 N^{-1}\sum_{j=1}^{N}\frac{\Delta^2\bar{E}_j}{\Delta\bar{B}^2}, \tag{2.28}$$

where μ_B is the Bohr magneton. The factor 2 in the second expression denotes the spin degeneracy of each level. The contributions to the sum in (2.28) are computed separately for each manifold of different symmetry (parity for $\sigma \neq 1$, k value for $\sigma = 1$). In this way, singularities of $\Delta^2\bar{E}_j/\Delta\bar{B}^2$ at true crossings are removed. $\Delta\bar{B} = 5.0 \times 10^{-2}$ is chosen here, which induces variations of eigenvalues $\{\bar{E}_j\}$ within their reliable decimal places. In fig. 2.8, the negative of χ is shown as a function of $\bar{B}$ for the case of $N = 100$ occupied levels, where approximately 50 levels are of even parity and the remaining ones are of odd parity.

For $\sigma = 1$, χ is found to retain the essential features of Landau diamagnetism in two-dimensional systems: $-\chi$ takes the largest value in the vicinity of $\bar{B} = 0$, and decreases monotonically with increasing B. Note that the χ determined by the contribution from the even-parity manifold alone exhibits almost identical results. Let us re-examine fig. 2.7(*a*). All crossings appearing in this figure are true crossings between levels with different k. Therefore, each energy level shows a smooth variation with $\bar{B}$ of positive curvature $\Delta^2\bar{E}_j/\Delta\bar{B}^2 > 0$, which decreases with increasing B. The characteristics of the diamagnetic susceptibility for $\sigma = 1$ shown in fig. 2.8 are thus well explained by the regular behavior of the spectrum which is a direct consequence of the integrability of this case.

For $\sigma = 0.5$, in contrast, χ shows remarkably different features. The value of $-\chi$ is greatly reduced at $\bar{B} = 0$ as compared with the Landau value, but increases on average with B, recovering the value for $\sigma = 1$

only for $\bar{E}_{100}/\bar{B}^2 \leq \rho_{\min}$, i.e., for $\bar{B} \geq 50$. This increase is accompanied by large fluctuations and anomalous dips (spikes), some of them even associated with positive values of χ. (For the spikes, see the comments below.) These features can be traced back to the behavior of the spectrum, which shows a multitude of avoided crossings (AC) in each of the two manifolds, see fig. 2.7(*b*). The rapid variation of the two levels with $\bar{B}$ near an AC gives rise to anomalous contributions of $\Delta^2\bar{E}_j/\Delta\bar{B}^2$ of opposite signs. If the AC is narrow and lies below ε_F, the two contributions cancel in (2.28), but most ACs have widths of order of or greater than their mutual distance, such that their effects overlap. This leads to a rather flat

Fig. 2.8 Negative of diamagnetic susceptibility χ as a function of $\bar{B}$ for $N = 100$. Circle and square symbols indicate $\sigma = 1$ and 0.5 cases, respectively. Heavy symbols and lines denote combined contributions from both even- and odd-parity manifolds, while fine counterparts denote contributions from the even-parity manifold alone with energies below ε_F. In figs. 2.8 and 2.10, χ is scaled by $(2mL^2/\hbar^2)\mu_B^2$.

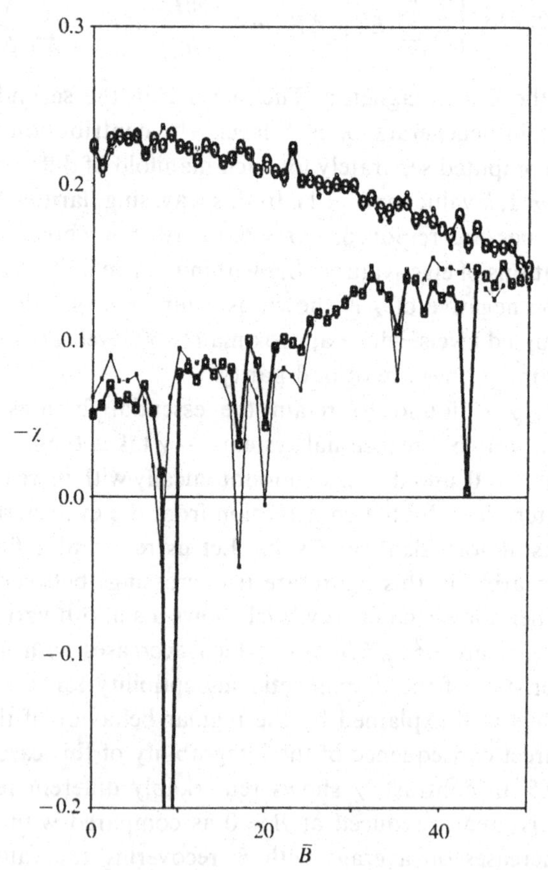

variation of each level with $\bar{B}$, with a greatly reduced average curvature and large fluctuations due to the nonuniform distribution of ACs. With increasing B most ACs become extremely narrow, and the B dependence of the levels approaches that for $\sigma = 1$. Anomalous dips in $-\chi$ occur for B values where ε_F lies within a gap of an AC. (From an experimental point of view, dips of this kind may be more or less suppressed by possible extrinsic disorder such as impurities.) Thus, the anomalous features of the diamagnetic susceptibility for $\sigma = 0.5$ shown in fig. 2.8 reflect the effects of level repulsion and avoided crossings typical for a nonintegrable system. Our additional data indicate a point worth noting: the reduction of χ becomes more remarkable and its fluctuations become less pronounced with decreasing σ towards $\sigma = 0$, while the opposite tendency is found with increasing σ towards $\sigma = 1$.

The present findings cannot simply be interpreted in terms of the traditional concept of bulk states and edge states (Teller, 1931), since such distinction of states is not possible for the case $R/L \sim 1$ considered here. Figure 2.9 shows an example of wavefunctions of a pair of states at a typical avoided crossing indicated in fig. 2.7(*b*). They exhibit a complicated delocalized structure due to superposition of contributions from many k values. In fact, they can be attributed neither to bulk nor to edge states.

To examine the physical relevance of the characteristics in fig. 2.8, we proceed to study the N dependence of χ for $\bar{B} = 0$ and 25 in figs. 2.10(*a*) and 2.10(*b*), respectively. There exists a clear tendency that differences of χ between $\sigma = 1$ and 0.5 are maintained, see fig. 2.10(*a*), or even enhanced, see fig. 2.10(*b*), for increasing N as long as $\bar{E}_N/\bar{B}^2 \geq \rho_{\min}$ for the highest occupied levels in the system with $\sigma = 0.5$. This ensures that our findings may be applied to thin conducting disks containing a low-density two-dimensional electron gas. Let us consider, for example, the interface layer in GaAs–AlGaAs heterojunctions, for which current research concentrates on the quantum Hall effect. In these systems, the electron concentration n is very low, typically, $n \simeq 10^{12}\,\text{cm}^{-2}$. Noting that the Fermi wavenumber $k_F \simeq n^{1/2}$ in two dimensions, we have $k_F \simeq 10^6\,\text{cm}^{-1}$. For the above n, $N = 10^2$ corresponds to $L \simeq 10^{-5}$ cm. Then $\bar{E}_N \simeq 8.0 \times 10^2$ for $\sigma = 0.5$ and the interesting phenomenon of quantum chaos is seen for $0 < \bar{B} \leq 50$, i.e., for $0 < B \leq 2.5$ T. The de Broglie wavelength $\sim k_F^{-1}$ is large enough for the results to be insensitive to lattice discreteness both within and along the boundary of the interface layer. In ordinary metal disks, on the other hand, n can be as large as $10^{16}\,\text{cm}^{-2}$, which corresponds to $k_F \simeq 10^8\,\text{cm}^{-1}$. Then the de Broglie wavelength is comparable to the lattice constant and so the results will be more or less modified.

We should make one more comment regarding the experimental viability of our theoretical concerns: coupling of electrons with atomic-scale defects (e.g., impurities and imperfections) and with phonons at finite temperatures, which are inevitable in real materials, may yield additional avoided crossings in the electronic energy spectra. However, the avoided crossings of this kind constitute the background level for the observed diamagnetic susceptibility, which is insensitive to the geometry of billiards. So our predictions concerning deterministic chaos can still be verified in real experiments for small two-dimensional devices so long as both the temperatures and the concentrations of atomic-scale defects are low enough.

To summarize, chaotic dynamics in the nonintegrable elliptic billiard in a uniform magnetic field is found to induce a remarkable reduction in and large fluctuations of the diamagnetic susceptibility, whereas the integrable circular billiard yields results close to Landau diamagnetism

Fig. 2.9 Wavefunctions $|\Psi|$ at the avoided crossing indicated by an arrow in fig. 2.7(*b*), $\sigma = 0.5$ and $\bar{B} = 10$: (*a*) $\bar{E}_{32} = 284.0230$; (*b*) $\bar{E}_{33} = 285.1459$.

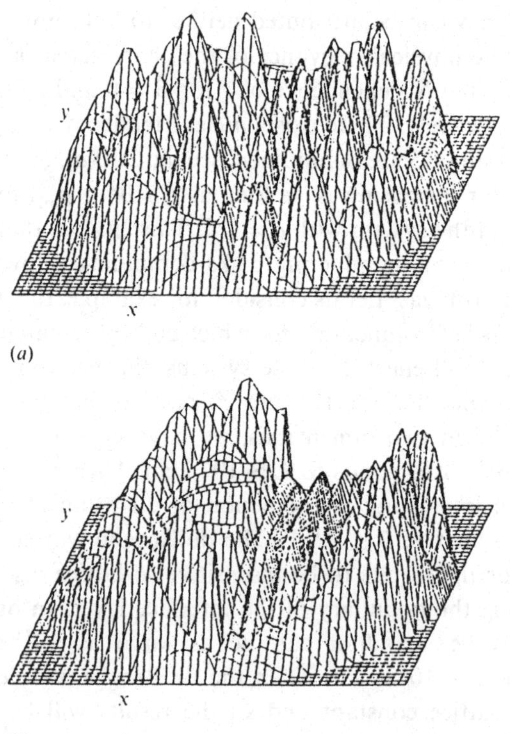

in two dimensions. We thereby have a subtle technique for catching the finger-prints of quantum chaos in experimentally accessible macroscopic quantum observables.

2.4 Open stadium billiards and quantum transport

Owing to recent progress in fabricating small electronic devices, a new target has become available for scientific investigations, i.e., a family of mesoscopic conducting disks with their linear dimension much smaller than the electronic mean free path ($\lesssim 10\ \mu$m) and comparable to the Fermi wavelength ($\simeq 100$ nm). (For a review, see Beenakker and van Houten (1991).) The free-electron gas picture holds good because there is a very small concentration (about $10^{12}\ \text{cm}^{-2}$) of electrons. In those disks, ballistic rather than diffusive motions of electrons are operative. Nonlinear dynamics of electrons in such small confining regions therefore plays a vital role, as recognized in sections 2.2 and 2.3. Current experiments, however, are almost all concerned with open systems and chaotic scattering is anticipated to have a crucial effect on ballistic quantum transport, e.g., normal resistance and Hall resistance. So, on moving from

Fig. 2.10 Plot of $-\chi$ as a function of N ($1 \leq N \leq 100$): (*a*) $\bar{B} = 0$; (*b*) $\bar{B} = 25$. Broken and solid lines correspond to $\sigma = 1$ and 0.5, respectively. The meaning of heavy and fine lines is the same as in fig. 2.8.

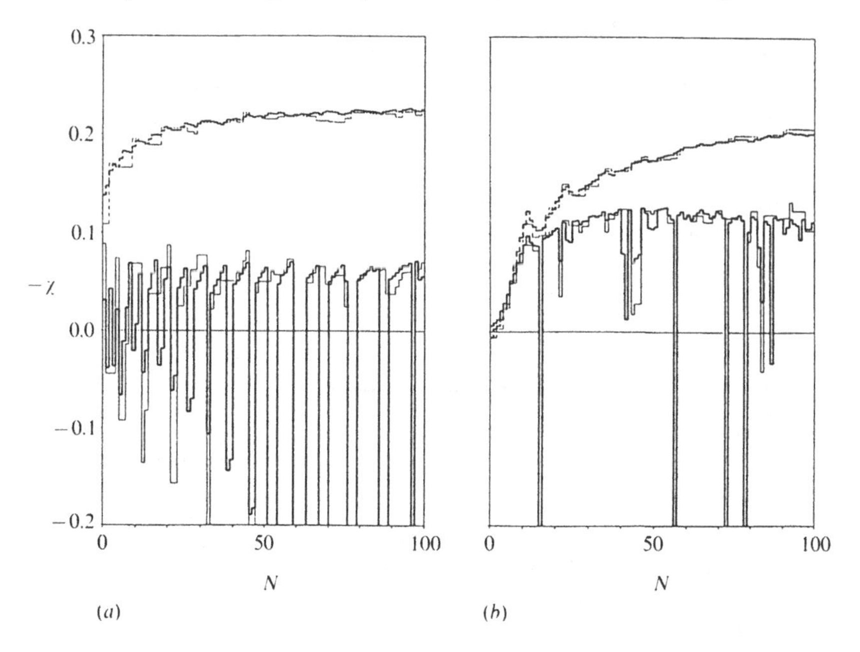

closed to open billiards, we should henceforth go beyond investigations limited to energy-level structures proper to closed systems.

In this section we shall choose concave and, in particular, stadium billiards (Bunimovich, 1974) as a model of mesoscopic disk. The magnetic field will be suppressed here for simplicity. Although there are some theoretical works on quantum transport in billiard systems (Roukes and Alerhand, 1990; Jalabert *et al.*, 1990; Baranger *et al.*, 1991), most of them are concerned with convex billiards (i.e., repellers or hyperbolic systems), where a clear relationship is available between escape rate and Lyapunov exponents. By contrast, we shall concentrate on open concave billiards bridging the integrable circle and the fully chaotic stadium. The stadium billiard belongs to the K systems and, together with the kicked rotator, constitutes a prototype of conservative chaotic systems. Quantum-mechanical studies of this system have sometimes had great impact on the field of quantum chaos (McDonald and Kaufman, 1979; Heller, 1984). Nevertheless, with a few exceptions, little attention has been paid to corresponding studies on its open-system version. In particular, differences between quantum transport in circle and stadium billiards are far from obvious.

Open concave billiards can be constructed by making a pair of holes at the boundary of the billiard and then by joining the semi-infinite conducting leads to the holes. As shown in fig. 2.11, this open system is characterized by a, l and d for semicircle radius, line segment half-length and hole width at $x = \pm a'$, respectively. Note that $a' = l + (a^2 - d^2/4)^{1/2}$. While the aspect ratio $\zeta \equiv l/a$ is tunable, the area of the billiard $\mathscr{A}(=\pi a^2 + 4al)$ is kept fixed and all lengths are scaled by $\mathscr{A}^{1/2}$. Here we shall concentrate on a weakly opened situation where ample fluctu-

Fig. 2.11 Open stadium billiard. + and $*$ are origins of (x, y) and s coordinates, respectively.

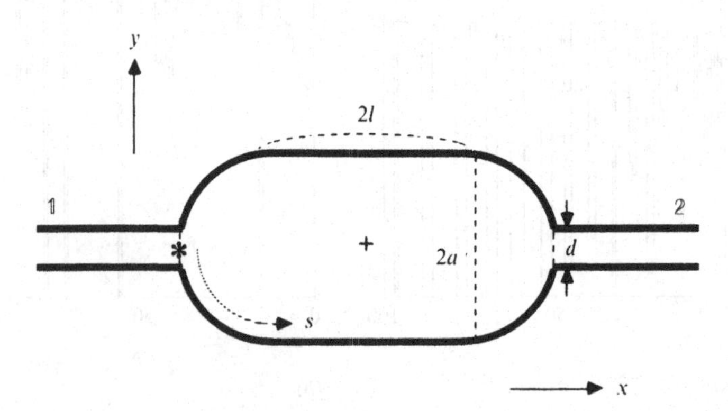

ation properties are anticipated in quantum transport. For convenience, $d/\mathscr{A}^{1/2} = 0.0935$ is chosen below. Edges of leads 1 and 2 are assumed located at left and right holes, respectively. We shall call the region lying inside the billiard and satisfying $|x| < a'$ the 'cavity region'. Chaotic scattering of electrons in this region will control conductance. The method used to analyze open systems greatly differs from the method in the previous section. Besides the boundary condition at hard walls, matching conditions at holes should be imposed such that an S-matrix is obtained. The S-matrix is a function of d, ζ and the Fermi energy $\varepsilon_F = (\hbar k_F)^2/2m$ of incoming electrons. In the context of nonlinear dynamics, the aspect ratio is the dominant parameter imposing chaos on the S-matrix, and our major interest lies in this effect. Once the S-matrix is available, we can calculate transmission and reflection coefficients, which in turn give rise to the conductance or resistance by application of Landauer's formula. The description below is based on the most recent work of Nakamura and Ishio (1992).

Before entering into details of quantum systems, let us review the features of classical billiards. For convenience, circle ($\zeta = 0$) and stadium ($\zeta = 1$) will be abbreviated to C and Sd, respectively. Figure 2.12 shows typical orbits of electrons in closed versions of C and Sd billiards. The nonergodic (C) versus ergodic (Sd) and periodic (C) versus erratic (Sd) behaviors can be recognized. It is also noteworthy that the C billiard has a caustic ensuring the stability of periodic orbits. Figure 2.13 shows the ζ dependence of K entropy (maximum Lyapunov exponent) calculated by Benettin and Strelcyn (1978). Except for the integrable limit $\zeta = 0$, K entropy is always positive, so stadium billiards with $\zeta > 0$ are called the K system. Figure 2.13 also indicates that the Sd billiard ($\zeta = 1$) is most chaotic.

In the open-system version of billiards, we can propose a new concept, namely the dwelling time τ during which an incoming electron stays in the cavity region. In figs. 2.14 and 2.15, τ is shown as a function of the initial location of the electron with given injection angle θ_0 at hole 1. For the Sd billiard, this dwelling-time spectrum exhibits fine comb-like structures consisting of multiple scales. Figure 2.14 demonstrates interesting self-similar or fractal structures. For the C billiard, neither fine structures nor self-similarity can be perceived (see fig. 2.15).

To see the quantum analog of these classical issues, we proceed to solve the Schrödinger equation for open systems, as given by

$$-\frac{\hbar^2}{2m}\nabla^2\Psi = \frac{\hbar^2}{2m}k_F^2\Psi. \tag{2.29}$$

Fig. 2.12 Orbits in closed billiards: (*a*) circle; (*b*) stadium ($\zeta = 1$).

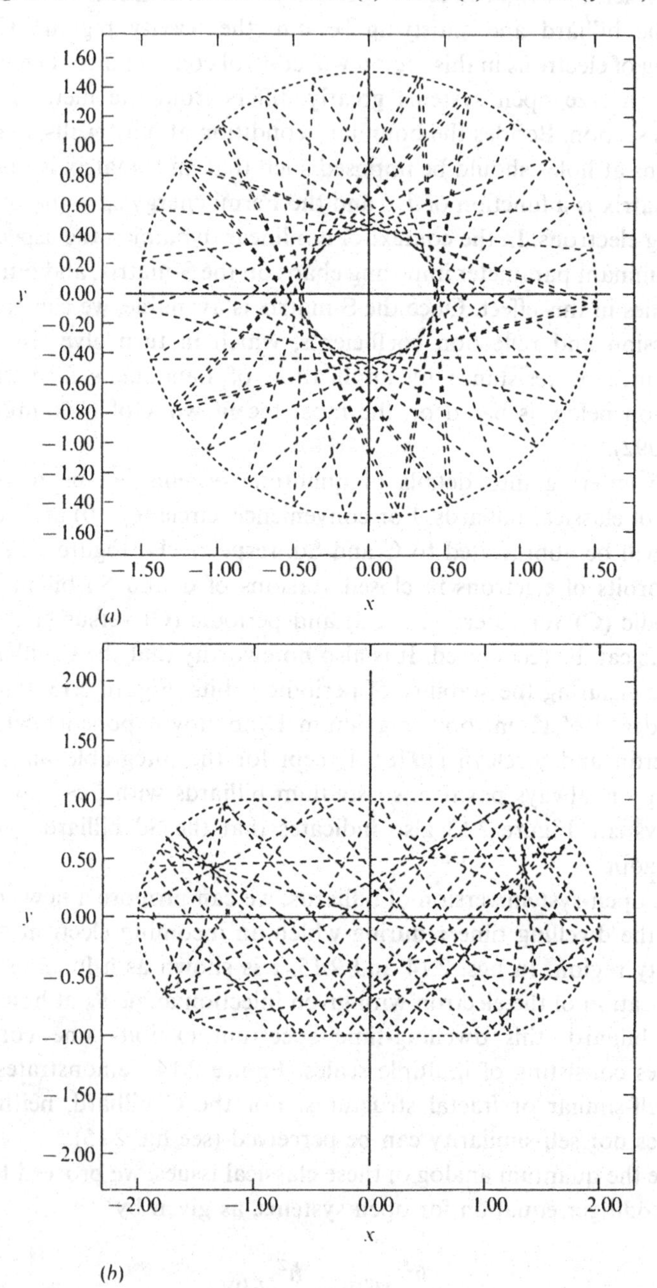

Let us define the wavevector $\mathbf{k} = (k_m, m\pi/d)$ at the leads, where $m\pi/d$ ($m = 1, 2, \ldots$) and $k_m = [k_F^2 - (m\pi/d)^2]^{1/2}$ are the transverse and longitudinal components, respectively. Channels (i.e., modes) m for $m \leq N$ and for $m > N$ with $N = k_F d/\pi$ correspond to propagating and evanescent waves, respectively.

Suppose a propagating wave incoming through lead 1 to have a mode $\mathbf{k} = (k_n, n\pi/d)$. The solution satisfying (2.29) is as follows. Inside the leads i ($=1, 2$), the solution is given by

$$\Psi_{\text{out}}^{(1)}(x, y; n) = e^{ik_n(x+a')}\phi_n(y) + \sum_{m=1}^{M} S_{mn}^{(1)}\, e^{-ik_m(x+a')}\phi_m(y) \qquad (2.30a)$$

and

$$\Psi_{\text{out}}^{(2)}(x, y; n) = \sum_{m=1}^{M} S_{mn}^{(2)}\, e^{ik_m(x-a')}\phi_m(y) \qquad (2.30b)$$

with $M > N$. In (2.30), $\{S_{mn}^{(i)}\}$ constitute the S-matrix, and $\{\phi_m\}$ defined by

$$\phi_m(y) = (2/d)^{1/2} \sin[(m\pi/d)(y + d/2)] \qquad \text{for } |y| \leq d/2 \qquad (2.31)$$

are transverse components of wavefunctions. Inside the billiard, on the other hand, a general solution of (2.29) is given as a superposition of plane waves by (e.g., Avishai and Band, 1987, 1989)

$$\Psi_{\text{in}}(x, y) = (2/d)^{1/2} \int_0^{2\pi} c_n(\theta)\, e^{i(p_x(\theta)x + p_y(\theta)y)}\, d\theta \qquad (2.32)$$

Fig. 2.13 The ζ dependence of K entropy. (Courtesy of G. Benettin and J. M. Strelcyn.)

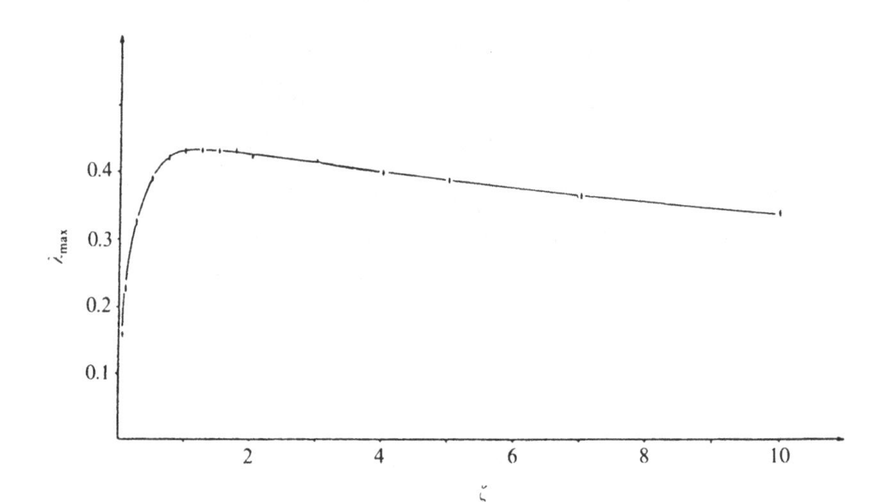

Fig. 2.14 Dwelling time (τ) spectrum for open stadium billiard ($\zeta = 1$). Abscissa denotes initial location of electron at hole 1. $\theta_0 = 0.993$ rad: (*a*) $-d/2 < y < d/2$; (*b*) partial magnification of (*a*).

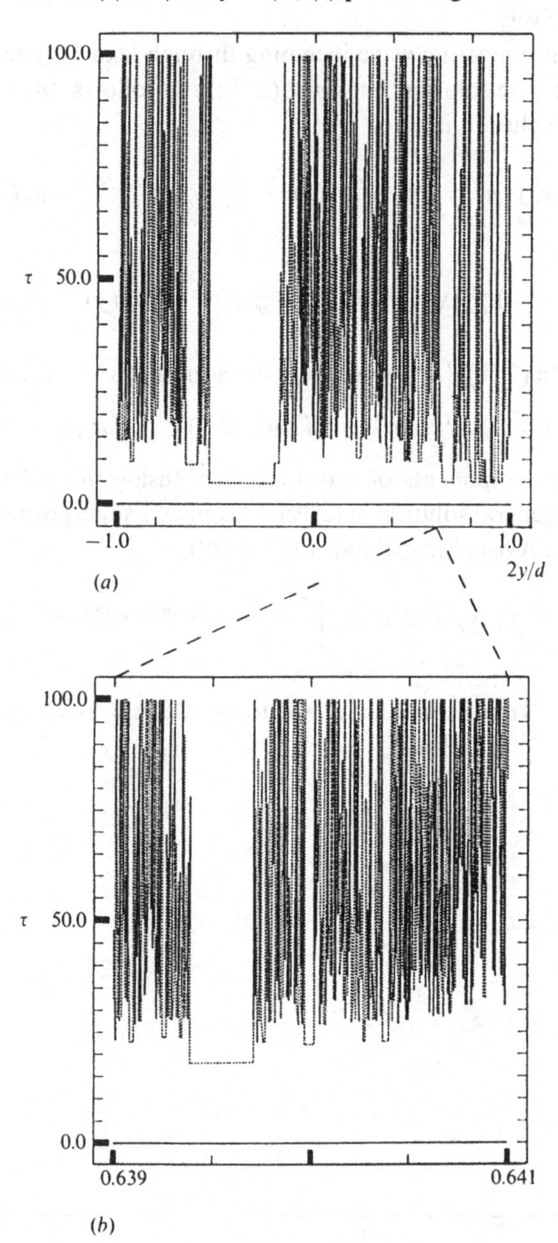

with

$$p_x(\theta) = k_F \cos\theta,$$
$$p_y(\theta) = k_F \sin\theta. \tag{2.33}$$

Boundary conditions are imposed as follows: (i) Ψ_{in} should satisfy the Dirichlet boundary condition at hard walls so that

$$\Psi_{in}(x, y) = 0; \tag{2.34}$$

(ii) (2.30) and (2.32) should match at the holes: the continuity and smoothness conditions are given respectively by

$$\Psi_{out}^{(i)}|_{x=a_i} = \Psi_{in}|_{x=a_i} \qquad (i = 1, 2) \tag{2.35}$$

and

$$\frac{d}{dx}\Psi_{out}^{(i)}|_{x=a_i} = \frac{d}{dx}\Psi_{in}|_{x=a_i} \qquad (i = 1, 2), \tag{2.36}$$

where $a_i = (-1)^i a'$. To summarize, we should solve (2.34) to (2.36) simultaneously to obtain $\{c_n\}$, and $\{S_{mn}^{(i)}\}$. Because of the mirror symmetry about the x axis, the solution will have the same parity as the incoming mode n. Therefore we can reduce (2.32) to a desymmetrized form

Fig. 2.15 Same as in fig. 2.14 but for open circle.

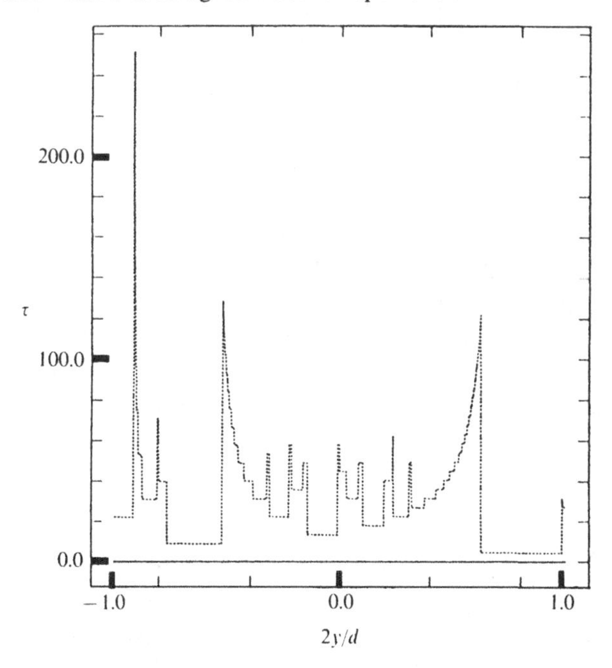

$$\Psi_{\text{in}}(x, y) = (2/d)^{1/2} \int_0^{\pi} c_n(\theta)\{e^{ip_y(\theta)y} - (-1)^n e^{-ip_y(\theta)y}\} e^{ip_x(\theta)x} d\theta. \quad (2.32')$$

Exploiting the orthonormality of $\{\phi_m(y)\}$ at each hole and taking into account the symmetry described above, the smoothness condition (2.36) can be rewritten as

$$\begin{aligned} S_{mn}^{(1)} &= \delta_{mn} - \int_0^{\pi} d\theta\, c_n(\theta) g_m(p_y(\theta))\, e^{-ip_x(\theta)a'}(p_x(\theta)/k_m), \\ S_{mn}^{(2)} &= \int_0^{\pi} d\theta\, c_n(\theta) g_m(p_y(\theta))\, e^{ip_x(\theta)a'}(p_x(\theta)/k_m), \end{aligned} \quad (2.37a)$$

with

$$g_m(p_y(\theta)) = 8m\pi[(p_y(\theta)d)^2 - (m\pi)^2]^{-1} \times \begin{cases} -\cos(p_y(\theta)d/2) & \text{for odd } m, n \\ i \sin(p_y(\theta)d/2) & \text{for even } m, n. \end{cases} \quad (2.37b)$$

The remaining conditions (2.34) and (2.35), on the other hand, have appearances that are mutually quite different, leading to ill-convergence in numerical computations. To overcome this difficulty, it is advantageous for us to move from real space to momentum space (Nakamura and Ishio, 1992). Let us assign a new coordinate s to positions on the hypothetical closed boundary of the cavity region. We define s as measured along the boundary anticlockwise from the origin at the center of hole 1 (see fig. 2.11). Any boundary function $F(s)$ has the common period $\Lambda(\sigma)$, i.e., perimeter length for the above closed boundary, so that it can be expanded as

$$F(s) = \sum_{l=-\infty}^{\infty} F_l\, e^{i(2\pi ls/\Lambda)}. \quad (2.38)$$

A set of equalities between corresponding Fourier coefficients of boundary functions $\Psi_{\text{out}}(s)$ and $\Psi_{\text{in}}(s)$ can substitute for (2.34) and (2.35). In our systems, independent equalities with $l = 0, 1, 2, \ldots, L$ and those with $l = 1, 2, \ldots, L$ will be chosen for even- and odd-parity solutions, respectively.

By rewriting boundary values (x, y) in terms of s, $\Psi_{\text{out}}(s)$ and $\Psi_{\text{in}}(s)$ are constructed from $\{\Psi_{\text{out}}^{(i)}(x, y)\}$ and $\Psi_{\text{in}}(x, y)$, respectively. For instance,

$$\Psi_{\text{out}}(s) = \begin{cases} \sum_{m=1}^{M}{}^{*} \{\delta_{mn}\phi_m(-s) + S_{mn}^{(1)}\phi_m(-s) & \text{for } |s| \le d/2 \\ \sum_{m=1}^{M}{}^{*} S_{mn}^{(2)}\phi_m\left(s - \dfrac{\Lambda}{2}\right) & \text{for } \left|s - \dfrac{\Lambda}{2}\right| \le d/2 \\ 0 & \text{otherwise,} \end{cases} \quad (2.39)$$

where $\sum_m^*$ implies a summation over m with the same parity as n. The Fourier coefficients of (2.39) are

$$\Psi_{\text{out},l} = \Lambda^{-1} \int_{-\Lambda/2}^{\Lambda/2} \mathrm{d}s\, \Psi_{\text{out}}(s)\, \mathrm{e}^{-\mathrm{i}(2\pi l s/\Lambda)}$$

$$= \Lambda^{-1}(2/d)^{1/2} \sum_{m=1}^{M}{}^{*} \{\delta_{mn} f_{ml} + (S_{mn}^{(1)} + (-1)^{l+m+1} S_{mn}^{(2)}) f_{ml}\} \quad (2.40a)$$

with

$$f_{ml} = \frac{2\pi m d}{(2\pi l d/\Lambda)^2 - (m\pi)^2} \times \begin{cases} -\cos(\pi l d/\Lambda) & \text{for odd } m \\ \mathrm{i}\sin(\pi l d/\Lambda) & \text{for even } m. \end{cases} \quad (2.40b)$$

Using (2.37) in (2.40), we have

$$\Psi_{\text{out},l} = \Lambda^{-1}(2/d)^{1/2} \left\{ 2 f_{nl} - \sum_{m=1}^{M}{}^{*} f_{ml} \int_0^{\pi} \mathrm{d}\theta\, c_n(\theta) g_m(p_y(\theta)) \right.$$

$$\left. \times [\mathrm{e}^{-\mathrm{i}p_x(\theta)a'} + (-1)^{l+m} \mathrm{e}^{\mathrm{i}p_x(\theta)a'}](p_x(\theta)/k_m) \right\}. \quad (2.41)$$

In the same way, Fourier coefficients of $\Psi_{\text{in}}(s)$ are given by

$$\Psi_{\text{in},l} = \Lambda^{-1}(2/d)^{1/2} \int_0^{2\pi} \mathrm{d}\theta\, c_n(\theta) \int_{-\Lambda/2}^{\Lambda/2} \mathrm{d}s\, [\mathrm{e}^{\mathrm{i}p_y(\theta)y} - (-1)^n \mathrm{e}^{-\mathrm{i}p_y(\theta)y}]$$

$$\times \mathrm{e}^{\mathrm{i}p_x(\theta)x}\, \mathrm{e}^{-\mathrm{i}(2\pi l s/\Lambda)}. \quad (2.42)$$

From the set of equalities $\{\Psi_{\text{out},l} = \Psi_{\text{in},l}\}$, we have

$$\int_0^{2\pi} \mathrm{d}\theta\, \mathbf{A}^{(n)} \mathbf{c}_n = \mathbf{b}^{(n)} \quad (2.43a)$$

with

$$A_{l\theta}^{(n)} = \sum_{m=1}^{M}{}^{*} g_m(p_y(\theta)) f_{ml} \times [\mathrm{e}^{-\mathrm{i}p_x(\theta)a'} + (-1)^{l+m} \mathrm{e}^{\mathrm{i}p_x(\theta)a'}](p_x(\theta)/k_m)$$

$$+ \int_{-\Lambda/2}^{\Lambda/2} \mathrm{d}s\, [\mathrm{e}^{\mathrm{i}p_y(\theta)y} - (-1)^n \mathrm{e}^{-\mathrm{i}p_y(\theta)y}] \times \mathrm{e}^{\mathrm{i}p_x(\theta)x}\, \mathrm{e}^{-\mathrm{i}(2\pi l s/\Lambda)} \quad (2.43b)$$

and

$$b_l^{(n)} = 2 f_{nl} \quad (2.43c)$$

where x and y are functions of s.

In our computations, integration with respect to θ in (2.43) is discretized so that $\mathbf{A}^{(n)}$ should be a square matrix. Using in (2.37) the solution $\mathbf{c}_n$ for (2.43), the S-matrix can be calculated in terms of submatrices $\mathbf{r}$ and $\mathbf{t}$ as

$$\mathbf{S} = \begin{pmatrix} \mathbf{r} & \mathbf{t} \\ \mathbf{t} & \mathbf{r} \end{pmatrix},$$

where $r_{mn} = (k_m/k_n)^{1/2} S_{mn}^{(1)}$ and $t_{mn} = (k_m/k_n)^{1/2} S_{mn}^{(2)}$ with $m, n = 1, 2, \ldots, N$ are flux-normalized reflection and transmission amplitudes, respectively. Since all matrix elements between different parities are vanishing, each of $\mathbf{r}$, $\mathbf{t}$ and $\mathbf{S}$ is decomposed into desymmetrized block matrices. The validity of the numerical solution for (2.37) is checked by using the unitarity of $\mathbf{S}$. The conductance is evaluated by $G = \sum_{n=1}^{N} G^{(n)}$ with $G^{(n)} = (2e^2/h) \sum_{1 \le m \le N} |t_{mn}|^2$ for each mode of incident wave.

Figures 2.16 and 2.17 show typical wavefunction features for the $n = 1$ mode. Series of mountains at lead 1 are due to the interference between incident and reflected waves. In the C billiard, $|\Psi|^2$ has a quantum analog of a caustic in the open system (fig. 2.16(*a*)) and other regular patterns with partially broken symmetry. In the Sd billiard, in contrast, $|\Psi|^2$ shows irregular patterns (fig. 2.17) with no remnant of circular symmetry. Both C and Sd billiards, despite the evident distinction between their wavefunction features, show equally an extreme sensitivity of the outgoing wave amplitude at lead 2 to k_F in the neighborhood of resonances, indicating the complicated conductance fluctuations in weakly opened cases.

Figure 2.18 shows the conductance $G^{(1)}$ as a function of k_F ($k_F d/\pi \le 3.2$). Because of the limitation in space, figures for $G^{(2)}$ and $G^{(3)}$ are omitted. Note that new modes appear at $k_F d/\pi = n$ with $n = 1, 2, \ldots$ and that G and $G^{(1)}$ are identical for $k_F d/\pi \le 2$. Both billiard types commonly exhibit very noisy fluctuations. These anomalous fluctuations, reminiscent of the universal conductance fluctuations in dirty metals, are a typical feature of weakly open systems, irrespective of their integrability or nonintegrability, so long as S-matrix poles are well concentrated. (In our treatment both C and Sd billiards have equal area $\mathscr{A}$ for their cavity region, ensuring almost equal concentration of poles.) Even in magnified views in fig. 2.18, it is hard to distinguish fluctuation features between C and Sd billiards. When a smoothed version $\mathscr{G}(k_F)$ is constructed from $G^{(1)}(k_F)$, however, astonishing differences appear: $\mathscr{G}$ shows in the C billiard case short-period ($T \simeq 0.25$) and small-amplitude ($A = 0.1$–0.2) oscillations around some plateau value (fig. 2.18(*a*)), whereas in the Sd billiard case long-period ($T \simeq 1.0$) and large-amplitude ($A \gtrsim 0.3$) smooth oscillations are found (fig. 2.18(*b*)). Rigorously speaking, the latter oscillations seem to show a slight irregularity but their comparatively monotonic behavior is

Fig. 2.16 Wavefunctions $|\Psi|^2$ for $n = 1$ mode in open circle: (*a*) $k_F d/\pi = 1.152$; (*b*) $k_F d/\pi = 1.273$.

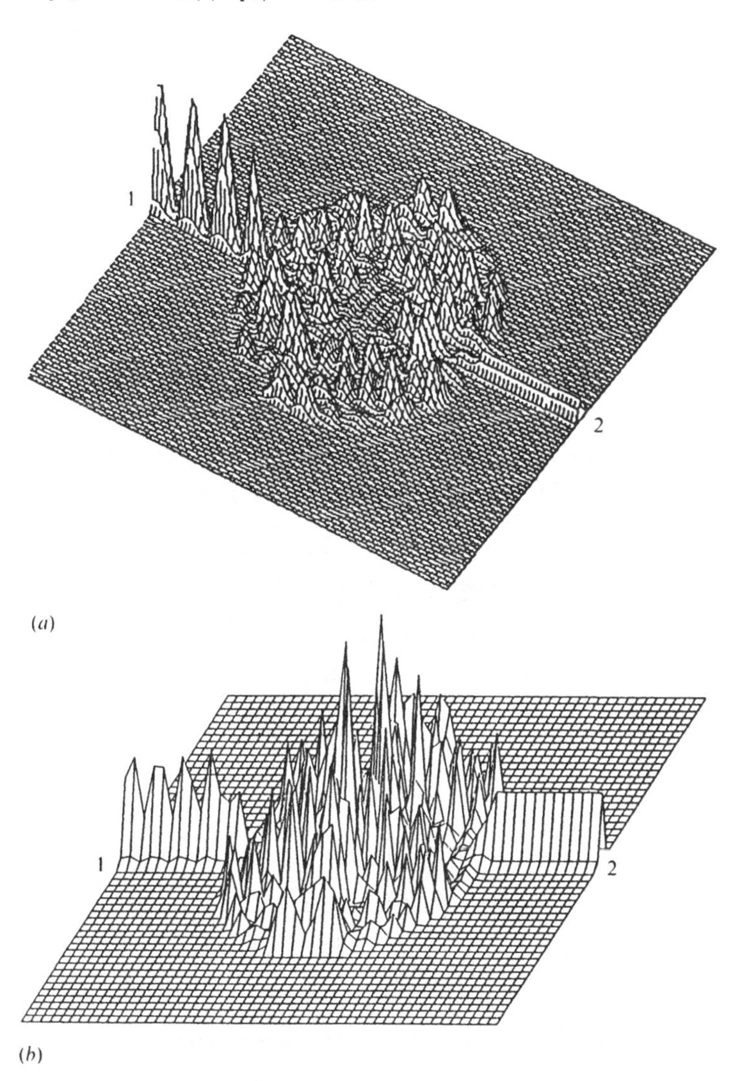

evident. Smoothing is done here by averaging $G^{(1)}(k_F)$ over each interval $(\Delta k_F)d/\pi = 0.08$. Successive intervals are chosen by shifting the preceding one by $(\delta k_F)d/\pi = 0.0008$. (We have confirmed that increasing or decreasing the above interval by $\pm 50\%$ does not alter the qualitative difference in $\mathscr{G}$ between C and Sd billiards.) The strong sensitivity of $\mathscr{G}$ to k_F in C billiards is responsible for the structural instability of classical phase space against a small change of energy in the integrable system. In contrast, the

Fig. 2.17 Same as in fig. 2.16 but for open stadium ($\zeta = 1$).

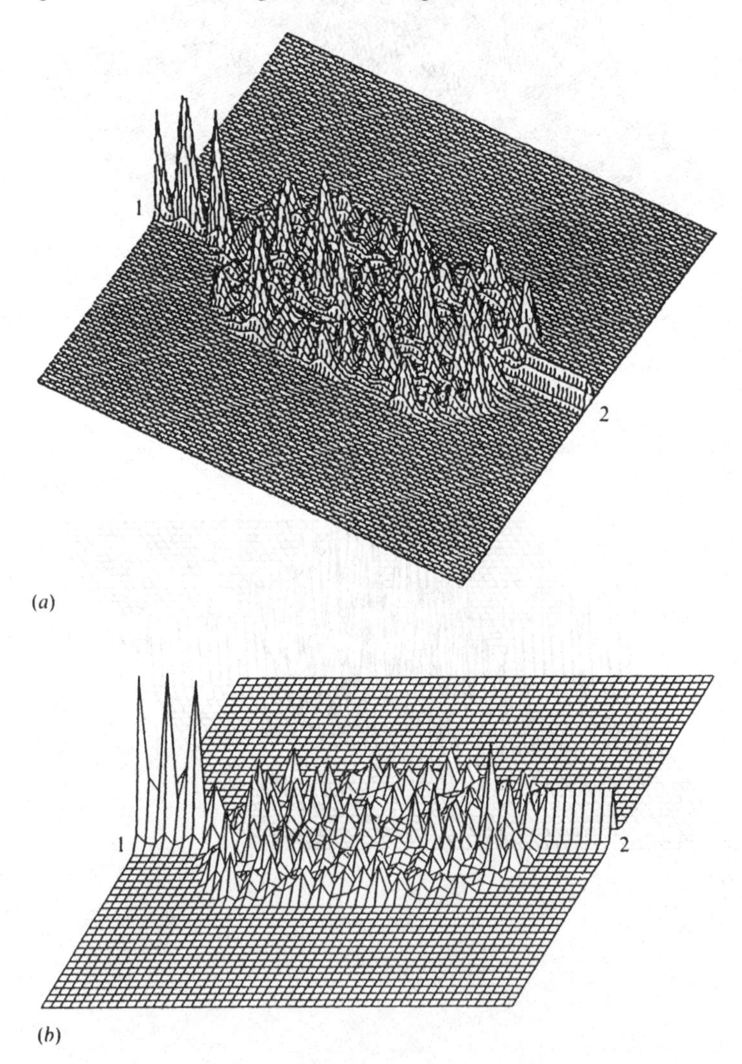

weak sensitivity of $\mathscr{G}$ to k_F in Sd billiards reflects the stability of classical ergodic phase space. Recall the analogous argument on closed quantum billiards in the previous section. We should note that while $G^{(n)}$ with $n \geq 2$ is also found to show the pronounced features above, the total conductance G (not shown here), even after smoothing, shows less distinct contrasts because of incoherence among oscillations in $G^{(n)}$ and $G^{(m)}$ for $n \neq m$.

To characterize the fluctuations, we have calculated the wavenumber autocorrelation function $\Gamma(\kappa) \equiv \langle \delta G^{(1)}(k_F - \kappa/2)\, \delta G^{(1)}(k_F + \kappa/2)\rangle_{k_F}$ with $\delta G^{(1)} \equiv G^{(1)} - \langle G^{(1)}\rangle_{k_F}$, whose definition is consistent with the standard correlation function studied in turbulence theory when the local average $\langle \cdots \rangle_{k_F}$ is interpreted as an ensemble average. (The average is now taken over $1.094 \leq k_F d/\pi \leq 1.903$.) $\Gamma(\kappa)$ in fig. 2.19 shows distinctive short-range behavior: (1) correlation as a whole is stronger in Sd than in C billiards; (2) the central peak at the origin has almost the same width for both Sd

Fig. 2.18 Conductance $G^{(1)}(k_F)$ for $k_F d/\pi \leq 3.2$. Dotted line represents smoothed conductance $\mathscr{G}$: (*a*) circle; (*b*) stadium ($\zeta = 1$). Insets: partial magnification of a range $1.4 \leq k_F d/\pi \leq 1.6$.

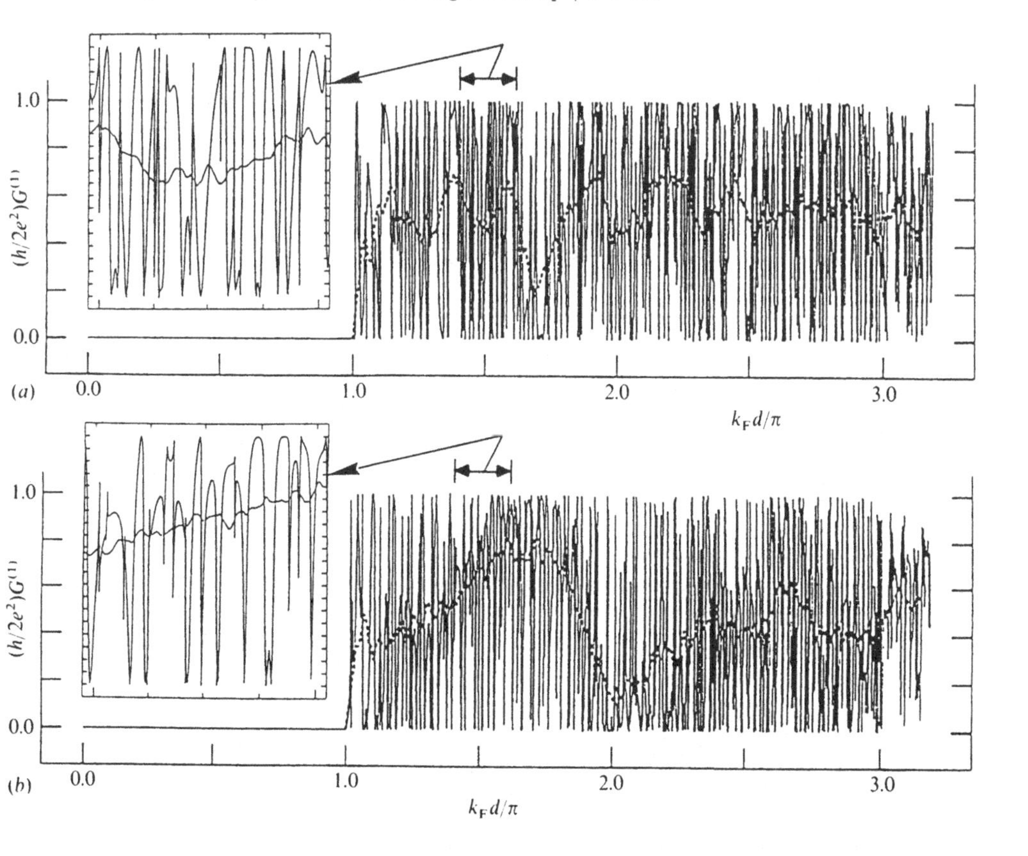

and C billiards. The result (1) provides a logical foundation for the monotonic structure of $\mathscr{G}(k_F)$ for Sd billiards in fig. 2.18(*b*). The result (2) indicates that both C and Sd billiards are commonly accompanied by Ericson-type fluctuations, necessitating enrichment of insight into the relationship between fluctuations in random matrix theory and nonlinear dynamics. The strong and weak short-range correlations in quantum transport remind us of Wigner and Poisson level-spacing distributions in closed systems, respectively. The contrasts between C and Sd billiards prove increasingly remarkable when $d/\mathscr{A}^{1/2}$ is decreased. For a continuous change of ζ from 0 through 1, the plateau becomes broken and conversion from short- to long-period and from small- to large-amplitude oscillations occurs. Further, in a strongly opened situation, when $d/\mathscr{A}^{1/2}$ is increased, conductance becomes quantized accompanied by a small number of intermittent dips due to rare resonances. Thus, conductance in weakly opened concave billiards has generic anomalous fluctuations controlled by the integrability or nonintegrability of the underlying classical dynamics.

Quantum-mechanical studies on closed stadium billiards have often been epoch-making: the first assignment of energy-level statistics to that of GOE was attempted by McDonald and Kaufman (1979); the discovery of periodic-orbit scars by Heller (1984) gave a pictorial foundation to Gutzwiller's semiclassical quantization. The study in this section has partly uncovered the nature of open stadium billiards for the first time.

Fig. 2.19 Wavenumber autocorrelation function $\Gamma(\kappa)$. Solid and dotted lines correspond to C and Sd billiards, respectively.

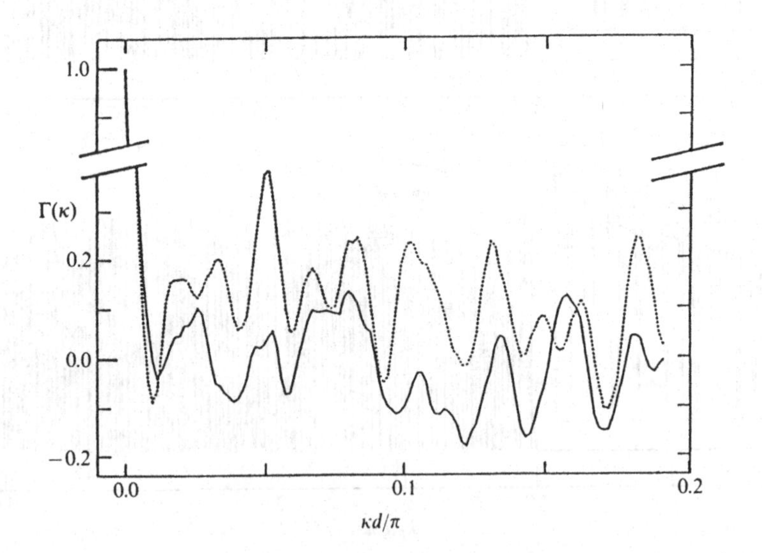

In traditional solid-state physics, conductance fluctuation in diffusive regimes in dirty metals has been studied extensively, in which extrinsic sources of randomness such as random potentials and impurities play a major role. In the present treatment, deterministic chaos is shown to give rise to analogous fluctuations in ballistic regimes in clean conducting disks in heterojunctions of compound semiconductors. The treatment developed here will be a vehicle for studying more complicated open billiards with four or more junctions and/or magnetic fields. As mentioned in chapter 1, quantum transport in mesoscopic disks also possesses great potential for unresolvable conflicts between experiments and quantum theory. So, these phenomena might be comparable to the blackbody radiation widely debated around the end of the last century.

3
Quantum chaos in spin systems

In this chapter, we provide two examples of quantum chaos in magnetic systems where only the spin degree of freedom plays an essential role. In the first half of the chapter, we study a nonintegrable triangular three-spin cluster with antiferromagnetic exchange coupling, and analyze the correspondence between chaotic dynamics and parameter-dependent semiclassical energy spectra. The curvatures for individual energy strings have fluctuations characterized by a scaling law in the semiclassically-large spin region. The scaling exponent is suggested to be a promising indicator of quantum chaos.

In the remaining half of this chapter, our investigation will move to quantum dynamics of a periodically pulsed single-spin system. We show quantum analogs of Kolmogorov–Arnold–Moser tori and chaos in the semiclassical regime. Wavefunctions (quasienergy states) in a spin-coherent state representation are characterized in terms of fractal dimensions. The time evolution of wavefunctions is quite remarkable: after a period of stretching- and folding-type diffusion, wavefunction patterns show ergodic and nonergodic features possessed by the underlying classical dynamics. Local dimensions (i.e., singularities) of their probability density functions maintain enhanced fluctuations when the pulse strength lies in a transitional region leading to global chaos.

3.1 Quantum chaos in an antiferromagnetic spin cluster: introduction

In the Monte Carlo simulation of lattice spin systems, extrinsic randomness (i.e., thermal noise) plays a major role, smearing out a deterministic aspect of the system. On the other hand, despite the accumulation of studies on the nonlinear dynamics of a variety of integrable and nonintegrable models, little attention has been given to irregular or chaotic dynamics in lattice spin systems – either classical or quantum. Rather, activity has concentrated on exceptional soliton limits for spin systems. These include several variants of the classical ferromagnetic Heisenberg chain ($\mathscr{J} > 0$)

$$\mathscr{H} = \mathscr{J} \sum_{i=1}^{\infty} (\mathbf{S}_i \cdot \mathbf{S}_{i+1} - \sigma S_i^z S_{i+1}^z). \tag{3.1}$$

For instance, in the continuum limit $\sigma = ga^2/2$ and

$$\mathbf{S}_{i+1} = \mathbf{S}_i + a \frac{\mathrm{d}\mathbf{S}_i}{\mathrm{d}x} + \tfrac{1}{2}a^2 \frac{\mathrm{d}^2\mathbf{S}_i}{\mathrm{d}x^2} + \cdots,$$

with the lattice spacing $a \to 0$, we have (Sklyanin, 1979)

$$\mathscr{H} \to \mathscr{H}_\mathrm{c} = \frac{\mathscr{J}}{2} \int \mathrm{d}x \left[\left(\frac{\mathrm{d}\mathbf{S}}{\mathrm{d}x} \right)^2 + 2g(S^z)^2 \right]. \tag{3.2}$$

It has been shown that the anisotropic Landau–Lifshitz equation for spin dynamics following from (3.2) is gauge-equivalent to the completely integrable continuum cubic nonlinear Schrödinger equation (Nakamura and Sasada, 1982). Again, the Lax pair structure associated with this continuum system can be discretized while maintaining integrability. This discrete Lax pair derives from a particular 'lattice log' extension of (3.1) with $\sigma = 0$:

$$\begin{aligned} \mathscr{H}_\mathrm{L} &= -\mathscr{J} \sum_i \ln(1 + \mathbf{S}_i \cdot \mathbf{S}_{i+1}) \\ &= -\mathscr{J} \sum_i [\mathbf{S}_i \cdot \mathbf{S}_{i+1} - \tfrac{1}{2}(\mathbf{S}_i \cdot \mathbf{S}_{i+1})^2 + \cdots]. \end{aligned} \tag{3.3}$$

Here the equation of motion is gauge equivalent to a completely integrable discrete nonlinear Schrödinger equation. As a final example, we note that most spin-$\frac{1}{2}$ quantum chains are completely integrable, e.g., by Bethe-ansatz or equivalently quantum-soliton techniques (Faddeev and Takhtajan, 1987).

The above examples, however, are, in a wider sense, exceptional. In this chapter we begin to investigate classical and quantum spin systems more generally, where we can continuously tune the degree of nonintegrability. For example, consider the XXZ antiferromagnetic chain, i.e., equation (3.1) with $\mathscr{J} < 0$ and $0 \leq \sigma \leq 1$. This Hamiltonian is completely integrable for $S = \frac{1}{2}$. However, there is no indication of integrability for $S > \frac{1}{2}$, including the classical limit $S \to \infty$. Here we limit ourselves to a three-spin triangular lattice with antiferromagnetic coupling, i.e.,

$$H = J \sum_{i=1}^{3} (\mathbf{S}_i \cdot \mathbf{S}_{i+1} - \sigma S_i^z S_{i+1}^z), \tag{3.4}$$

with $J > 0$, $0 \leq \sigma \leq 1$, and periodic boundary conditions (see fig. 3.1). Thus the $S = \frac{1}{2}$ case is again completely integrable. We find strong numerical evidence for nonintegrability in both the classical limit and semiclassical (large S) regimes, except in the isotropic limit $\sigma = 0$. Consequently, this is a convenient system in which to study systematically quantum-irregular spectra as a function of both the nonintegrability (σ) and degree of quantization ($\hbar \sim S^{-1}$). Although the number of spins is small, we can exploit the advantages of being able to cover a large range of S and of being able to handle the symmetry in spin and real spaces to obtain conveniently spectra for large S. This kind of small-spin cluster may be directly relevant in real materials, e.g., trinuclear complexes. It also constitutes a cornerstone for further study of extended antiferromagnetic triangular lattices, e.g., $NaTiO_2$, $RbFeCl_3$.

The major convenience of our spin-cluster model is that we can readily vary both σ and S^{-1} to study (i) the intuitive correspondence between classical chaos and quantum irregular spectra in semiclassical regions, and (ii) global aspects of the quantum energy-level distribution as a function

Fig. 3.1 Triangular three-spin clusters; θ and ϕ denote polar and azimuthal angles, respectively.

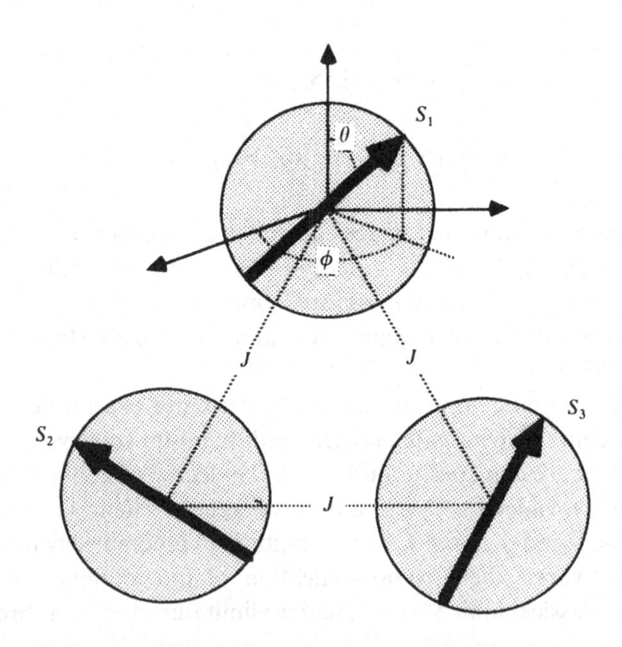

of both nonintegrability (σ) and degree of quantization (S^{-1}). In particular, we find strong evidence for scaling in the fluctuation of second-order differences of eigenvalues as $S \to \infty$, and quasifractal features in the level distribution for irregular spectral regimes.

Two further advantages of this spin system are worth noting. First, unlike cases of particle motion in finite potential wells or atomic motion in small molecules, we are able to examine the eigenvalue distribution over the total energy range without qualitative changes due to an 'escape' energy or quantum tunneling. This allows us to study a crossover energy between predominantly localized and predominantly extended states (reminiscent of an 'Anderson transition' in extrinsically disordered quantum models) over a wide energy range (especially as a function of σ). Second, chaotic and otherwise irregular behavior occurs at low energies, even including the vicinity of the ground state (see sections 3.2 and 3.3). This is in contrast to many models in which 'quantum chaos' has been examined, and is primarily due to antiferromagnetic spin coupling. In general, we can expect a reversal of energy regimes for corresponding ferromagnets with complex dynamics prevalently at high energies, which is somewhat less interesting, experimentally, for solid-state (e.g., magnetic) materials. We should also note that the triangular lattice with antiferromagnetic coupling is, as S decreases, 'frustrated' (i.e., there are competing interactions). This leads us to anticipate complexity at low energies such as the 'resonating-valence-bond' ground state (Anderson, 1973) that may be responsible for high-T_c superconductivity (Anderson, 1987).

Sections 3.2 through 3.5 are organized as follows. In section 3.2 we describe results for classical dynamics obtained from Poincaré surfaces of section. Sections 3.3 and 3.4 contain our results for eigenvalue distributions in semiclassical quantum cases (with $S \leq 36.5$). In particular, we discuss an apparent scaling law satisfied by the fluctuation of second-order differences of eigenvalues for increasing (large) S. We also suggest relationships between the scaling exponent, apparent fractal features of level distribution (as a function of σ), and a semiclassical analog of classical chaos. In section 3.5, we summarize our results and discuss future research motivated by them. The description in these four sections is based on work by Nakamura *et al.* (1985) and Nakamura and Bishop (1986).

3.2 Classical treatment

In this section we present results for the Hamiltonian dynamics following from (3.4) in the classical limit $S = \infty$. Then $\mathbf{S}_j$ is a three-component vector obeying the discrete Landau–Lifshitz equation of motion (without any dissipation)

$$\frac{d\mathbf{S}_j}{dt} = \{\mathbf{S}_j, H\} = \mathbf{S}_j \times (-\delta H/\delta \mathbf{S}_j). \tag{3.5}$$

Here (A, B) is a classical Poisson bracket:

$$\{A, B\} \equiv \sum_{\alpha\beta\gamma,j} \varepsilon_{\alpha\beta\gamma} \frac{\partial A}{\partial S_j^\alpha} \frac{\partial B}{\partial S_j^\beta} S_j^\gamma. \tag{3.6}$$

Since the magnitude of each spin is conserved, (3.5) describes dynamics in, at most, a six-dimensional phase space. Choosing $\mathbf{S}_j^2 = 1$, $\mathbf{S}_j$ may be written in terms of polar (θ_j) and azimuthal (ϕ_j) angles as

$$\mathbf{S}_j = (\sin\theta_j \cos\phi_j, \sin\theta_j \sin\phi_j, \cos\theta_j). \tag{3.7}$$

In terms of canonical variables $q_j = \phi_j$, $p_j = \cos\theta_j$, (3.5) can be rewritten as

$$\frac{dq_j}{dt} = \{q_j, H(\{q_j, p_j\})\} \tag{3.8a}$$

$$\frac{dp_j}{dt} = \{p_j, H(\{q_j, p_j\})\} \tag{3.8b}$$

with the conventional Poisson bracket

$$\{A, B\} \equiv \sum_j \left(\frac{\partial A}{\partial p_j} \frac{\partial B}{\partial q_j} - \frac{\partial B}{\partial p_j} \frac{\partial A}{\partial q_j} \right). \tag{3.9}$$

The present system therefore describes a Hamiltonian system with three degrees of freedom. Liouville's theorem (Arnold, 1978) indicates that such a system is completely integrable if three independent constants of motion exist.

The first integral of motion is the energy E. We introduce the total magnetization

$$\mathbf{T} = \sum_{j=1}^{3} \mathbf{S}_j. \tag{3.10}$$

For the isotropic limit $\sigma = 0$ in (3.4), both $\mathbf{T}^2$ and T^z are conserved, reflecting the SO(3) symmetry of the system. They are also mutually commuting. In total, we have three independent constants of motion, ensuring complete integrability of the system with $\sigma = 0$. For $\sigma \neq 0$, on the other hand, the symmetry is lowered from SO(3) to SO(2) and $\mathbf{T}^2$ is no longer a constant of motion. Consequently, the system with $\sigma \neq 0$ may be nonintegrable and chaotic. (For two-spin systems the integrability condition can be more easily found (Magyari *et al.*, 1987).) Promising tests for nonintegrability involve applying Ziglin's analysis (Ziglin, 1983a, 1983b) or Painlevé's test (Ablowitz *et al.*, 1980; Daniel *et al.*, 1992), but here we restrict ourselves to implications from Poincaré surfaces of section.

In the following classical treatment we set $J = 1$ (without loss of generality) and choose $\mathbf{S}_j^2 = 1$ and $T^z = 0$. This last condition is appropriate for antiferromagnetic ground-state ordering and limits our choice of initial data to the phase around that ground state. The ground state ($\sigma > 0$) is given by a 120° structure in the S^x–S^y plane: e.g., $\theta_1 = \theta_2 = \theta_3 = \pi/2$ and $\phi_1 = 0$, $\phi_2 = 2\pi/3$, $\phi_3 = 4\pi/3$ (see fig. 3.1). Initial data are then selected from

$$\begin{aligned} \theta_1^{(0)} &= \pi/2, \qquad \theta_2^{(0)} = \pi/2 + \beta, \qquad \theta_3^{(0)} = \pi/2 - \beta, \\ \phi_1^{(0)} &= \alpha, \qquad \phi_2^{(0)} = 2\pi/3, \qquad \phi_3^{(0)} = 4\pi/3, \end{aligned} \tag{3.11}$$

where α and β are arbitrary parameters denoting basal-plane disorder and off-plane angles, respectively. (Note from (3.11) that $T^z = 0$.) Here we have suppressed arbitrary additional constants common to ϕ_i ($i = 1, 2, 3$) coming from SO(2) symmetry of the system (3.4). The conserved energy E is easily determined from (3.11) in (3.4) as

$$E = -\cos\alpha\cos\beta - (1 - \sigma) - (\sigma - \tfrac{1}{2})\cos^2\beta. \tag{3.12}$$

The ground-state energy is then

$$E(\alpha = \beta = 0) = E_{\mathrm{G}} = -1.5.$$

The highest energy state, in which three spins align parallel in the x–y plane, also satisfies $T^z = 0$ and has energy $E_{\mathrm{H}} = 3$. For each fixed energy in the range $E_{\mathrm{G}} \leq E \leq E_{\mathrm{H}}$, several consistent sets of (α, β) are chosen and the resulting long-time dynamics are analyzed by Poincaré surfaces of section, specified by positive crossings of $dS_1^z/dt = 0$ – these three-dimensional sections are projected on to the S_1^x–S_1^y plane.

For illustration, we concentrate here on results for $\sigma = 0.3$. (Trends for $\sigma \to 0, 1$ are discussed at the end of this section.) Projected surfaces of section are given in figs. 3.2(*a*)–3.2(e). We see that, at high energies, most of the orbits are regular KAM tori, see fig. 3.2(*a*), whereas at low energies, irregular ergodic orbits dominate. An upper 'transition' energy is $E_{c_2} \simeq -0.9$ for the limited initial data considered; i.e., for $E \leq E_{c_2}$ KAM tori are successively destroyed – note from figs. 3.2(*b*)–3.2(c) that destruction of KAM-like tori at high energy is easiest for larger β; i.e., initial data with

Fig. 3.2 Projection onto S_1^x–S_1^y plane of Poincaré surface of section for $\sigma = 0.3$: (*a*) $E = -1.0$; (*b*) $E = -1.1$; (*c*) $E = -1.25$; (*d*) $E = -1.35$; (*e*) $E = -1.45$. Orbits with initial angles $\beta = 3n\pi/50$ ($n = 1$–5) are shown in each figure.

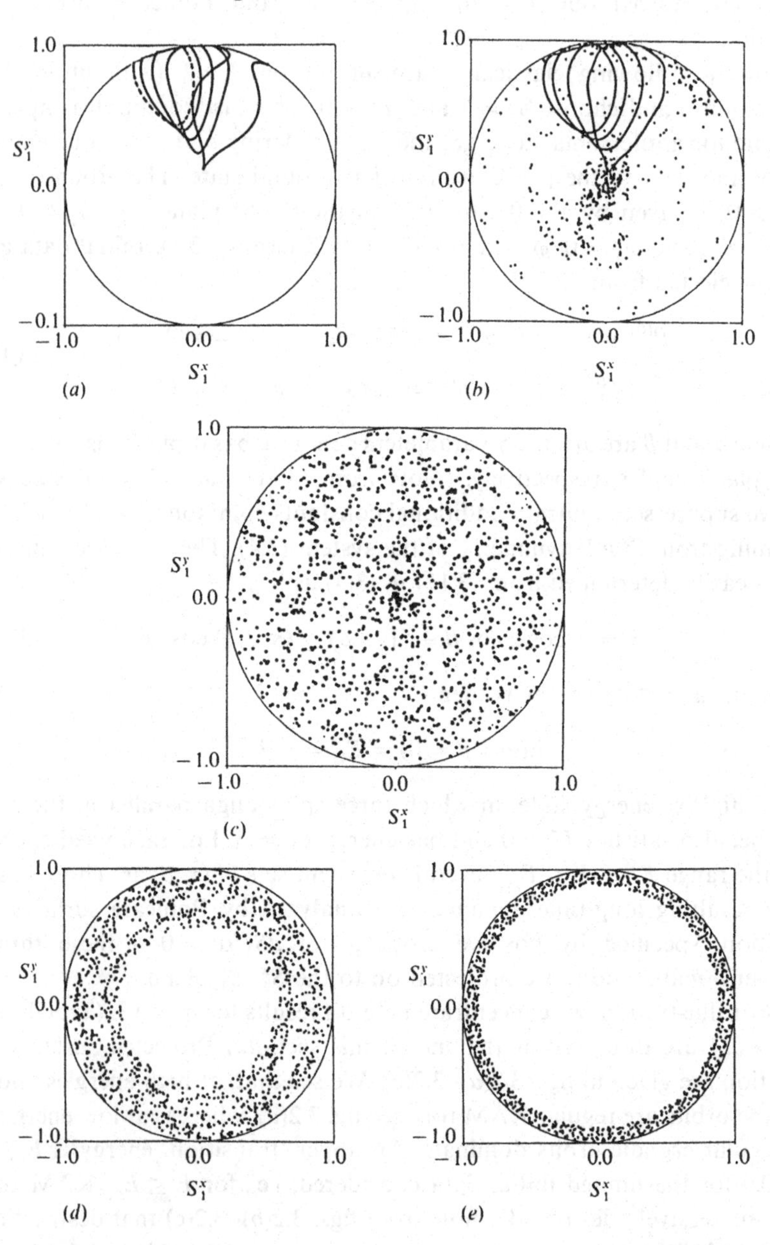

large off-plane angle leads to chaos most easily. The penetration of ergodic orbits into the torus region in fig. 3.2(*b*) does not imply Arnold diffusion (Lichtenberg and Lieberman, 1983). The penetration is due to an artifact caused by an inappropriate choice of cross-section. In fact, to see Arnold diffusion in our three-spin system, we must break the SO(2) symmetry, reducing the system to a fully anisotropic XYZ model.

As E is further lowered, accessible phase space is restricted to the annulus, see fig. 3.2(*d*). However, extensive chaos persists even close to the classical ground-state energy E_G, i.e., down to a lower 'transition' energy $E_{c_1} \simeq -1.4$. For $E_G \leq E \leq E_{c_1}$, only KAM orbits corresponding to spin-wave excitations survive, see fig. 3.2(*e*). Some narrow region at $E \simeq -1.25$ between E_{c_1} and E_{c_2} seems to admit only chaotic orbits, suppressing the coexistence of KAM tori.

Behavior for general σ ($0 \leq \sigma \leq 1$) is qualitatively similar to $\sigma = 0.3$: in the integrable isotropic limit $\sigma = 0$, only periodic orbits are observed with E_{c_1} and E_{c_2} merging to E_G. In the XY limit $\sigma = 1$, the width of the range in which to find chaos ($\Delta E = E_{c_2} - E_{c_1}$) is increased with both E_{c_1} and E_{c_2} showing upward shifts. Our more complete data indicate that as σ increases from 0 through 1, (i) both E_{c_1} and E_{c_2} move upwards and (ii) both the coexistence region ($E_{c_1} \leq E \leq E_{c_2}$) and a fully chaotic region inside it grow gradually. In the following sections, we study the quantum-mechanical counterparts of these findings.

3.3 Quantum treatment: construction of matrix elements

We now proceed to the quantum treatment, examining the distribution of eigenvalues and its σ and S dependence. Calculation of matrix elements of Hamiltonian (3.4) is facilitated by using the C_{3v} symmetry of the system and the planar SO(2) symmetry in spin space. The former symmetry is equivalent to the combined lattice translation ($\hat{T}$) and inversion ($\hat{I}$) symmetries. Basic 'kets' are classified by using a set of quantum numbers: wavenumber K ($=0, 2\pi/3, 4\pi/3$), parity P ($= +1, -1$), and total magnetization T^z (denoting integer, half-integer). An example of basic 'kets' with a given set of quantum numbers are

$$\begin{aligned}\psi_K^{\pm}(M_a, M_b, M_c) = \frac{1}{2^{1/2}}\bigg[&\frac{1}{3^{1/2}}(|SM_a\rangle_1 \times |SM_b\rangle_2 \times |SM_c\rangle_3 \\ &+ e^{-iK}|SM_b\rangle_1 \times |SM_c\rangle_2 \times |SM_a\rangle_3 \\ &+ e^{-2iK}|SM_c\rangle_1 \times |SM_a\rangle_2 \times |SM_b\rangle_3) \\ &\pm \frac{1}{3^{1/2}}(\text{cyclic in order of a, c and b})\bigg] \qquad (3.13)\end{aligned}$$

for nonidentical sets M_a, M_b and M_c. The second term in square brackets in (3.13) implies inversion of the first.

Compounding the constituent spins by using Racah coefficients is not appropriate in the present problem, where the total spin is not a good quantum number in general. Below, we give some details of how to construct basic kets and matrix elements.

Wavefunctions $|SM_a\rangle_1 \times |SM_b\rangle_2 \times |SM_c\rangle_3$ are abbreviated as $|M_a M_b M_c\rangle$ hereafter. The z component of the total magnetization is $T^z \equiv M_a + M_b + M_c$; without loss of generality we assume $M_a \le M_b \le M_c$. T^z is restricted to integer or half-integer values, but here we further restrict T^z to the $\frac{1}{2}$ or 0 manifolds which are favorable to the antiferromagnetic ground state. (Note that $T^z = 0$ is equivalent to the $T^z = 0$ constraint adopted in section 3.1.)

Since $M_{a,b,c} \in (-S, -S+1, \ldots, S-1, S)$, we have

$$-S \le M_a \le \tfrac{1}{3}T^z. \tag{3.14}$$

Similarly, we readily find

$$M_a \le M_b \le \tfrac{1}{2}T^z - \tfrac{1}{2}M_a \tag{3.15}$$

and

$$T^z - M_a - S \le M_b \le \tfrac{2}{3}T^z - M_a. \tag{3.16}$$

Inequalities (3.14) and (3.15) give

$$\mathrm{Max}\,(M_a, T^z - M_a - S) \le M_b \le \tfrac{1}{2}T^z - \tfrac{1}{2}M_a. \tag{3.17}$$

Also,

$$M_c = T^z - M_a - M_b. \tag{3.18}$$

Conditions (3.14), (3.17) and (3.18) fully constrain M_a, M_b and M_c.

Turning now to the C_{3v} space symmetry of the triangular lattice (symmetry $\hat{T} \times \hat{I}$, with $\hat{T}$ and $\hat{I}$ translation and inversion operations, respectively), we observe from elementary group-theory considerations that irreducible representations are classified by wavenumber K $(=0, 2\pi/3, 4\pi/3)$ and parity P $(=+1, -1)$. Basic 'kets' are then classified by the quantum numbers K, P and T^z. These 'kets' are classified into three types:

Type I: $M_a \ne M_b$, $M_b \ne M_c$, $M_c \ne M_a$. Now we have the form given in (3.13). Translation and inversion operators give

$$\hat{T}\psi_K^{\pm} = e^{iK}\psi_K^{\pm} \tag{3.19}$$

and

$$\hat{I}\psi_K^{\pm} = \pm\psi_K^{\pm}. \tag{3.20}$$

Here the choice of + or − corresponds to *gerade* or *ungerade* symmetry, respectively.

Type II: $M_a = M_b \neq M_c$ or $M_a \neq M_b = M_c$. Now

$$\psi_K = \frac{1}{3^{1/2}}(|M_a M_b M_c\rangle + e^{-iK}|M_b M_c M_a\rangle + e^{-2iK}|M_c M_a M_b\rangle). \qquad (3.21)$$

Here only *gerade* symmetry is admissible.

Type III: $M_a = M_b = M_c$. Now

$$\psi = |M_a M_b M_c\rangle, \qquad (3.22)$$

since only $K = 0$ and *gerade* symmetry is possible. Note also that restricting ourselves to $T^z = \frac{1}{2}$ makes type-III basis functions impossible in the three-spin system.

To construct the matrix elements for the quantum version of the Hamiltonian (3.4), we note that

$$S_i^x S_{i+1}^x + S_i^y S_{i+1}^y = \tfrac{1}{2}(S_i^+ S_{i+1}^- + S_i^- S_{i+1}^+) \qquad (3.23)$$

and

$$\langle M'|S^z|M\rangle = M\delta_{M',M}, \qquad (3.24a)$$

$$\langle M'|S^\pm|M\rangle = [(S \mp M)(S \pm M + 1)]^{1/2}\delta_{M',M\pm 1}, \qquad (3.24b)$$

where $S_j^\pm \equiv S_j^x \pm iS_j^y$. We also confine ourselves here to the manifold to $K = 0$, $P = +1$. The results below are already lengthy and extension to $K \neq 0$ is straightforward. Note that mixing between manifolds with different K is prohibited unless translational invariance is broken, e.g., by a spatially inhomogeneous magnetic field.

Consider a pair of $K = 0$ basis functions

$$\Phi_0(M_a M_b M_c) \equiv \Phi_0 = \frac{1}{3^{1/2}}(|M_a M_b M_c\rangle + |M_b M_c M_a\rangle + |M_c M_a M_b\rangle) \qquad (3.25a)$$

and

$$\Phi_0'(M_a' M_b' M_c') \equiv \Phi_0' = \frac{1}{3^{1/2}}(|M_a' M_b' M_c'\rangle + |M_b' M_c' M_a'\rangle + |M_c' M_a' M_b'\rangle). \qquad (3.25b)$$

Noting that, from (3.19), $\hat{T}\Phi_0 = \hat{T}^2\Phi_0 = \Phi_0$, we have

$$\begin{aligned}\langle\Phi_0'|H|\Phi_0\rangle &= \frac{1}{3^{1/2}}(\langle\Phi_0'|H|M_aM_bM_c\rangle + \langle\Phi_0'|H\hat{T}|M_aM_bM_c\rangle \\ &\quad + \langle\Phi_0'|H\hat{T}^2|M_aM_bM_c\rangle) \\ &= \langle M_a'M_b'M_c'|H|M_aM_bM_c\rangle + \langle M_b'M_c'M_a'|H|M_bM_cM_a\rangle \\ &\quad + \langle M_c'M_a'M_b'|H|M_cM_aM_b\rangle. \end{aligned} \tag{3.26}$$

To derive the final form in (3.26), we have used $H = \hat{T}H\hat{T}^{-1} = \hat{T}^2H\hat{T}^{-2}$. Noting that $H = \sum_{i=1}^{3} H_{i,i+1}$ with $H_{31} = \hat{T}H_{12}\hat{T}^{-1}$ and $H_{23} = \hat{T}^2H_{12}\hat{T}^{-2}$, three matrix elements in the final expression of (3.26) can be given in terms of matrix elements of H_{12}. For instance,

$$\begin{aligned}\langle M_a'M_b'M_c'|H|M_aM_bM_c\rangle &= \langle M_a'M_b'M_c'|H_{12}|M_aM_bM_c\rangle \\ &\quad + \langle M_b'M_c'M_a'|H_{12}|M_bM_cM_a\rangle \\ &\quad + \langle M_c'M_a'M_b'|H_{12}|M_cM_aM_b\rangle. \end{aligned} \tag{3.27}$$

Using (3.24) and (3.27) in (3.26), we arrive at

$$\begin{aligned}\langle\Phi_0'|H|\Phi_0\rangle = (&\{\delta_{M_c'M_c}\{\tfrac{1}{2}[(S-M_a)(S+M_a+1)(S+M_b)(S-M_b+1)]^{1/2} \\ &\times \delta_{M_a',M_a+1}\delta_{M_b',M_b-1} \\ &+ \tfrac{1}{2}[(S-M_b)(S+M_b+1)(S+M_a)(S-M_a+1)]^{1/2} \\ &\times \delta_{M_b',M_b+1}\delta_{M_a',M_a-1} + (1-\sigma)M_aM_b\delta_{M_a',M_a}\delta_{M_b',M_b}\}\} \\ &\times \{a\to b, b\to c, c\to a\} + \{a\to c, b\to a, c\to b\}) \\ &+ (M_a'\to M_b, M_b'\to M_c', M_c'\to M_a') \\ &+ (M_a'\to M_c', M_b'\to M_a', M_c'\to M_b'). \end{aligned} \tag{3.28}$$

Finally, we need to distinguish between types I and II.

(A) Type I × type I. Here,

$$\Psi_0^+ = \frac{1}{2^{1/2}}[\Phi_0(M_aM_bM_c) + \Phi_0(M_aM_cM_b)],$$

$$\Psi_0'^+ = \frac{1}{2^{1/2}}[\Phi_0'(M_a'M_b'M_c') + \Phi_0'(M_a'M_c'M_b')],$$

so that

$$\begin{aligned}\langle\Psi_0'^+|H|\Psi_0^+\rangle &= \langle\Phi_0(M_a'M_b'M_c')|H|\Phi_0(M_aM_bM_c)\rangle \\ &\quad + \langle\Phi_0'(M_a'M_c'M_b')|H|\Phi_0(M_aM_bM_c)\rangle, \end{aligned} \tag{3.29}$$

where we have used $H = \hat{I}H\hat{I}^{-1}$.

(B) Type I × type II and type II × type I. Here, for instance,

$$\langle \Psi_0'^{+}(M_a' M_b' M_c')|H|\Phi_0(M_a M_b M_c)\rangle$$

$$= \frac{1}{2^{1/2}} [\langle \Phi_0'(M_a' M_b' M_c')|H|\Phi_0(M_a M_b M_c)\rangle + \{\Phi_0'(M_a' M_c' M_b')|H|\Phi_0(M_a M_b M_c)\rangle], \tag{3.30}$$

(C) Type II × type II. Matrix elements in this case are already given in (3.28).

Thus we have obtained a block matrix specified by a definite set of quantum numbers ($K = 0$, $P = +1, T^z = \frac{1}{2}$). Diagonalization of this matrix will be done in cases $S = \frac{1}{2}, 4\frac{1}{2}, \ldots, 32\frac{1}{2}$ and $36\frac{1}{2}$. Computed energies are divided by $S(S+1)$ so that these scaled energies become densely populated in a classically admissible energy range ($-1.5 \le E \le 3.0$) for $S \gg 1$.

3.4 Irregular energy spectra

In fig. 3.3, 153 energy levels for $S = 16\frac{1}{2}$ are plotted as a function of σ. The energy spectra in fig. 3.3 are completely desymmetrized ($K = 0$, $P = +1, T^z = \frac{1}{2}$) except for the limit $\sigma = 0$. The degeneracy at $\sigma = 0$ is caused by SO(3) symmetry. Degenerate levels can be classified by the quantum number associated with $\mathbf{T}^2$. For $\sigma > 0$, however, each set of degenerate levels broadens and overlaps with adjacent sets. Since $\mathbf{T}^2$ is no longer a good quantum number, many avoided crossings appear (von Neumann and Wigner, 1929). For $\sigma = 0.3$, for instance, avoided crossings prevail in the range $E_{c_1}'(S) \le E \le E_{c_2}'(S)$ with $E_{c_1}'(S) \simeq -1.25$ and $E_{c_2}'(S) \simeq -0.5$ for $S = 16\frac{1}{2}$. This fact resembles the result in the previous section, in which classical chaos can be found in a similar energy range to that above. It is therefore possible to assert that the presence of many avoided level crossings is a quantum-mechanical manifestation of classical chaos.

The classical and quantum threshold values, however, are far from identical: $E_{c_1}'(S) > E_{c_1}$ and $E_{c_2}'(S) > E_{c_2}$. This discrepancy can be observed throughout the range $0 < \sigma \le 1$ in fig. 3.3. To establish precisely the quantum and classical correspondence, therefore, more careful insight will be required.

When the spin magnitude is increased beyond $S = 16\frac{1}{2}$, the fluctuation of each energy string in energy spectra would be reinforced due to the growing number of avoided crossings. To characterize these fluctuations, the second-order differences $\Delta^2 E/\Delta\sigma^2$ may be defined for each energy string:

$$\Delta^2 E/\Delta\sigma^2 \equiv [E(\sigma + \Delta\sigma) - 2E(\sigma) + E(\sigma - \Delta\sigma)]/(\Delta\sigma)^2. \quad (3.31)$$

This is a local curvature of the σ-dependent quantum energy diagram and describes the sensitivity of eigenvalues to the nonintegrability parameter. Quantity (3.31) was originally proposed by Pomphrey (1974), but here we examine its statistical averages:

$$K(S) \equiv \langle \Delta^2 E/\Delta\sigma^2 \rangle, \quad (3.32)$$

where the mean $\langle \cdots \rangle$ implies the average over levels lying within an appropriate energy range at a fixed value σ.

Below, we choose $\Delta\sigma = 5 \times 10^{-4}$, which yields the variation of $E(\sigma)$ within reliable decimal scales. The S dependence of $K(S)$ is then presented. In fig. 3.4, the averages $\langle \cdots \rangle$ are taken within individual energy ranges: the lower-energy part ($E_G \leq E \leq 0.0$) is divided equally into six energy bins, while the higher-energy part ($0.0 \leq E \leq 3.0$) is divided equally into two bins. We find: (i) the histogram in each energy bin shows some

Fig. 3.3 The σ-dependent energy diagram for $S = 16\frac{1}{2}$. The right-hand plot is a partial magnification of the left. Throughout figs. 3.3–3.6, energy is scaled by $S(S + 1)$.

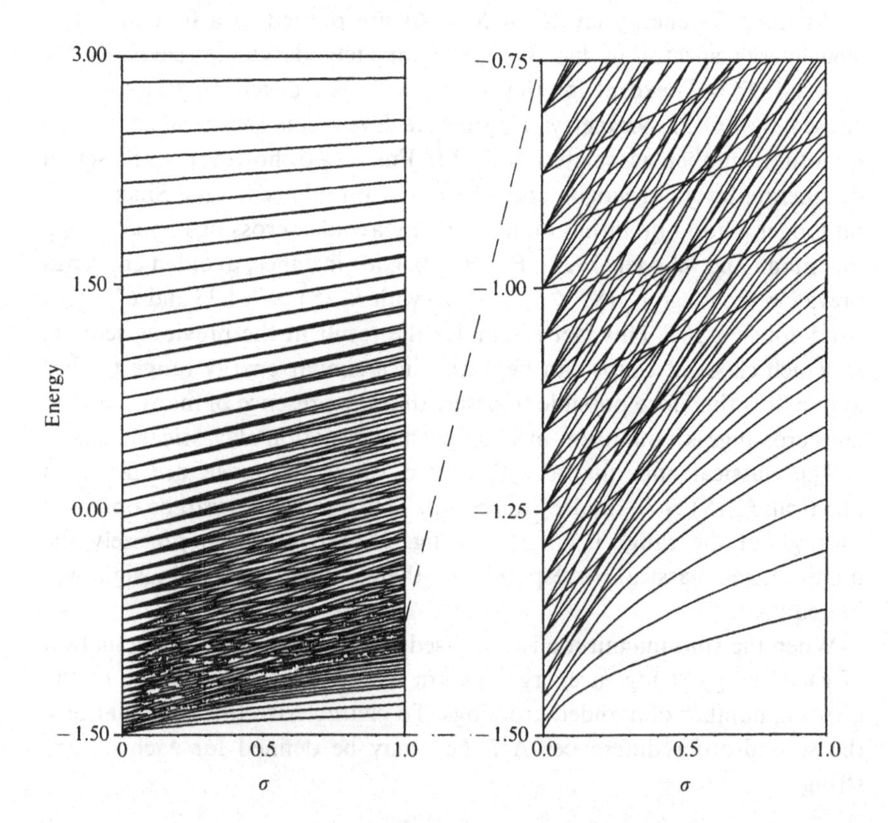

Fig. 3.4 S dependence of $K(S)$ in (3.32). Energy range is equally divided into six and two bins for $-1.5 \leq E \leq 0.0$ and $0.0 \leq E \leq 3.0$, respectively. In the histogram in each bin, values K(S) from the left correspond to $S = 8\frac{1}{2}, 12\frac{1}{2}, \ldots, 32\frac{1}{2}$: (*a*) $\sigma = 0$; (*b*) $\sigma = 0.3$; (*c*) $\sigma = 1.0$.

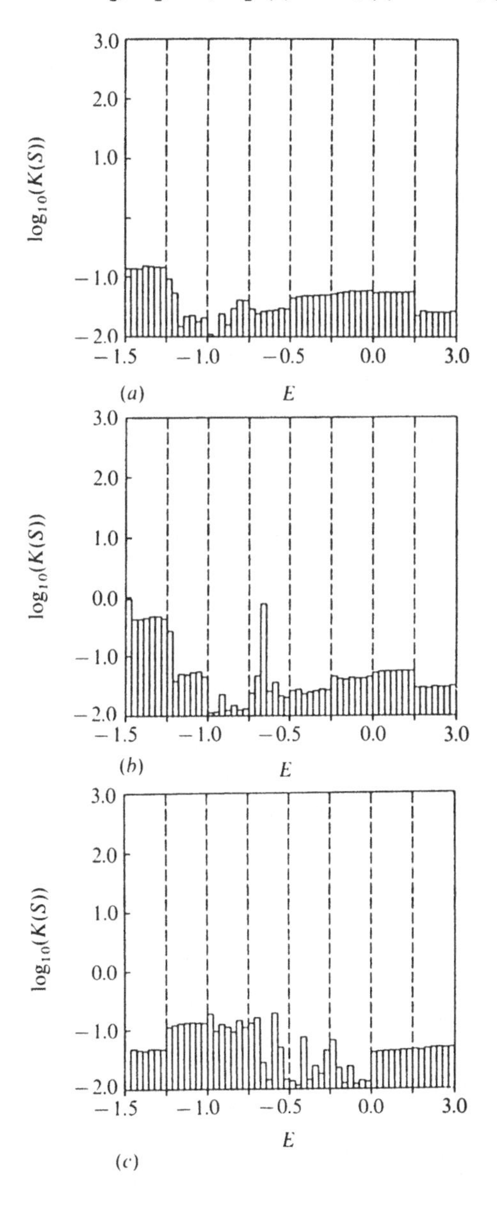

variations depending on the energy range; (ii) $K(S)$ exhibits, however, no significant σ and S dependence in large-S regions. The mean curvature $K(S)$ itself is therefore not a relevant quantity for characterizing quantum chaos.

The situation will be dramatically changed by introducing the fluctuation or standard deviation of curvatures:

$$G(S) \equiv \langle (\Delta^2 E/\Delta\sigma^2 - \langle \Delta^2 E/\Delta\sigma^2 \rangle)^2 \rangle, \tag{3.33}$$

where the meaning of $\langle \cdots \rangle$ is the same as in (3.32). The dependence of $G(S)$ on S is given in figs. 3.5(*a*)–(c). In the case of $\sigma = 0.3$ (see fig. 3.5(*b*)), $G(S)$ basically grows with increasing S in the ranges $-1.50 \lesssim E \lesssim -0.75$. For other energy ranges $-0.75 \lesssim E \lesssim 3.0$, increasing S keeps $G(S)$ almost constant; 'spikes' in the histograms are accidental, arising from sharp avoided crossings (see also section 2.3). More abundant data available indicate that this remarkable growth of $G(S)$ occurs precisely in the range $-1.40 \lesssim E \lesssim -0.90$. In the classical treatment for $\sigma = 0.3$, widespread chaos was observed only for $E_{c_1} \lesssim E \lesssim E_{c_2}$ with $E_{c_1} \simeq -1.4$ and $E_{c_2} \simeq -0.9$ for initial data restricted to the limited part of phase space. When σ is increased towards $\sigma = 1.0$ (see fig. 3.5(*c*)), the energy range where the growth of $G(S)$ can be expected, i.e., the 'quantal chaotic energy range', becomes broadened and accords with the classically chaotic regime. When σ is decreased towards $\sigma = 0.0$ (see fig. 3.5(*a*)), the quantal chaotic energy range becomes vanishing, in agreement with the classical result. Therefore, the semiclassical thresholds $E'_{c_1}(\infty)$ and $E'_{c_2}(\infty)$, between which the standard deviation of curvatures grows with increasing S, might be promising critical energies for characterizing the quantum version of classical chaos:

$$E'_{c_1}(\infty) \simeq E_{c_1}; \qquad E'_{c_2}(\infty) \simeq E_{c_2}. \tag{3.34}$$

The quantum thresholds $E'_{c_1}(S)$ and $E'_{c_2}(S)$ for a fixed value of S are anticipated to move towards $E'_{c_1}(\infty)$ and $E'_{c_2}(\infty)$, respectively, when S is increased.

The S dependence of $G(S)$ is shown on logarithmic scales in fig. 3.6, which suggests the existence of a scaling law of the form

$$G(\Lambda S)/G(S) = \Lambda^{\beta(\sigma)}, \tag{3.35}$$

where Λ is an arbitrary multiplication factor. The exponent $\beta(\sigma)$ is positive in the classically chaotic region (see lines b and c in figs. 3.6(*b*) and 3.6(*c*), respectively). On the other hand, it is almost zero in the classically regular as well as integrable ($\sigma = 0$) regions. The present scaling arguments provide a possibility of resolving the difficulty in elucidating the quantum version of classical chaos.

Fig. 3.5 S dependence of $G(S)$ in (3.33). Captions are the same as in fig. 3.4: (*a*) $\sigma = 0$; (*b*) $\sigma = 0.3$; (*c*) $\sigma = 1.0$.

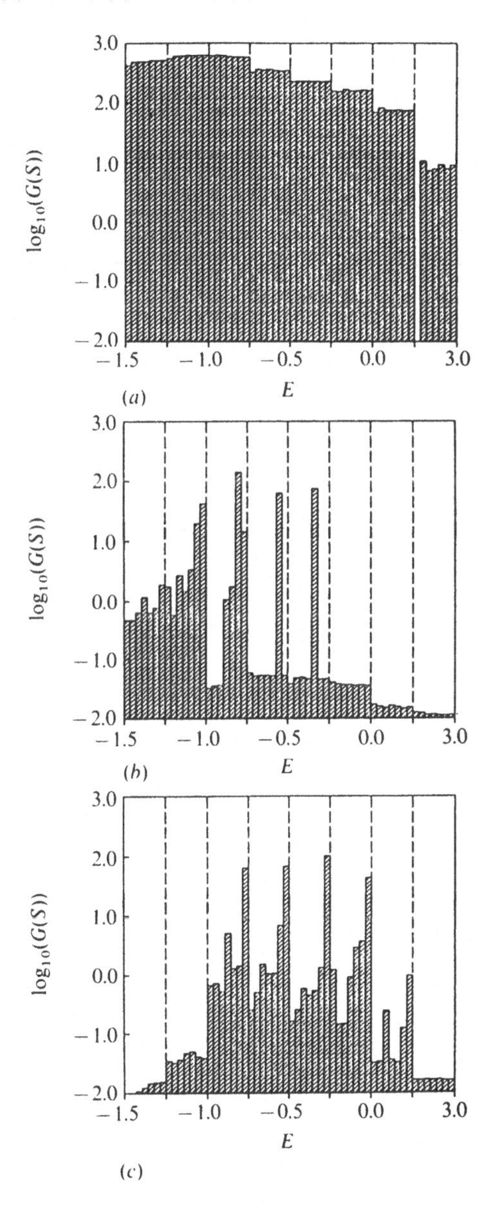

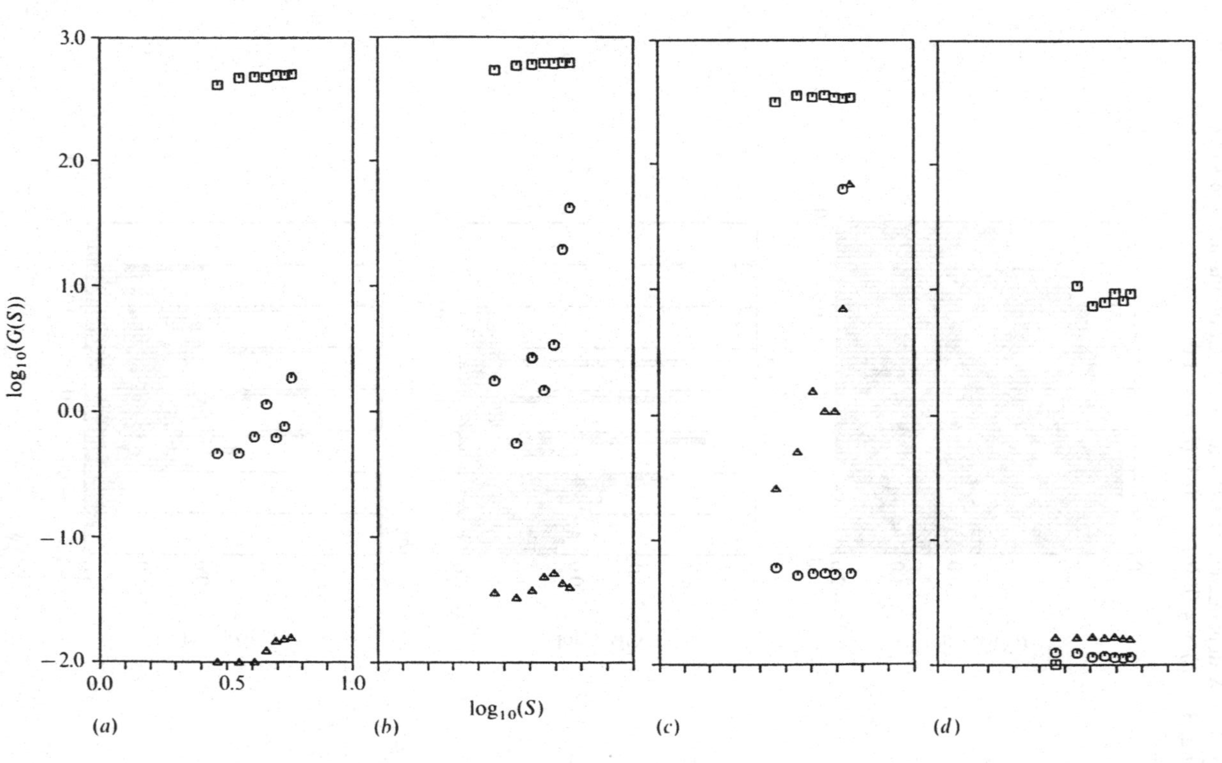

Fig. 3.6 S dependence of $G(S)$ in logarithmic scales. Lines a, b and c correspond to $\sigma = 0$ (□), $\sigma = 0.3$ (○), $\sigma = 1.0$ (△), respectively: (*a*) average over $-1.50 \leq E \leq -1.25$; (*b*) average over $-1.25 \leq E \leq -1.0$; (*c*) average over $-0.75 \leq E \leq -0.50$; (*d*) average over $1.50 \leq E \leq 3.0$.

3.5 New speculations

In sections 3.1–3.4, we have investigated the correspondence between classical and semiclassical chaos in a lattice spin system. In particular, we have demonstrated the nature of the quantum level distribution as a function of both nonintegrability (σ) and degree of quantization (S^{-1}). This information leads us to make the following new speculations. (i) Fractal features in the semiclassical eigenvalues distribution correspond to chaos in the classical limit. Here we have in mind fractal structure as a function of integrability σ. The fractal structure is a result of avoided level crossings, which are especially evident because of the densely packed levels as $S \to \infty$. (ii) A quantitative characterization of the fractal nature of the level distribution is given by a scaling exponent β in (3.35) for fluctuations of the second-order differences of eigenvalues, as the classical limit is approached ($S \to \infty$), where averages are taken over appropriate energy ranges. The cases $\beta = 0$ and $\beta > 0$ correspond to regular (as well as integrable) and chaotic regions in the underlying classical dynamics, respectively. Note that large curvature of energy levels as a function of σ is not in itself a criterion for a quantum analog of classical chaos. (iii) Apparent discrepancies between quantum irregular spectra and classical chaos have been debated in recent literature. These discussions have been based on transitions between localized and irregular spectra in models (e.g., Hénon–Heiles) where nonintegrability and $\hbar$ were fixed. As we have seen (section 3.4), the ability to study readily $\hbar \to 0$ (i.e., $S^{-1} \to 0$) in our model leads us to a new critical energy in general. We propose that $E'_{c_1}(S)$ and $E'_{c_2}(S)$ are natural criteria for transition energies between which the quantum analog of classical chaos appears. As $\hbar \to 0$ ($S^{-1} \to 0$), structure in the semiclassical level distribution evolves differently for integrable and nonintegrable cases.

The model and results presented here raise a number of questions and avenues for future exploration. As is well known, anomalous classical diffusion gives a clearer signature for the transition energy $E_c(\sigma)$ in the classical limit between strongly ergodic and KAM regimes. This point of view may connect naturally to the Anderson-localization-like transition in the quantum case. In the quantum situation, delocalization of localized wavefunctions associated with remnant periodic or KAM-like orbits arises from tunneling between overlapping wavefunction tails and leads to irregular spectra. Quantum diffusion should again be a sensitive diagnostic (as it is in the classical Anderson-localization problem where disorder is extrinsic). This, however, raises the question of evaluating both wave-functions and time dependence in the quantum case. We can pursue these properties using matrix-element input as derived in section 3.3. Clearly in regimes with regular spectra, the smallest level spacing will lead to a dominant exponential decay with time. However, in the presence of irregular spectra, multiple timescales, power law, etc., decay can be anticipated. Transport, power spectra and other quantum dynamic signatures will be much more powerful as well as of direct experimental relevance.

In this study, the number of levels available (703 for $S = 36\frac{1}{2}$) is sufficient to explore level-spacing distribution – we might expect a Poisson distribution (Berry and Tabor, 1977b) for $\sigma = 0$ and a Wigner distribution for $\sigma > 0$. However, increasing S or the number of spins will give us better statistics, so that we might better assess the validity of these distributions or suggested interpolations, e.g., that proposed by Brody *et al.* (1981). Increasing the size of the spin cluster could, of course, also improve statistics for small S.

It is clearly important to understand how 'universal' the behaviors we have found in the present model are. In addition to extending currently popular models to varying $\hbar$, we propose to study several other few-body models in which nonintegrability and degree of quantization can also be readily controlled. These include (i) other lattice-spin systems, e.g., a four-spin complex, antiferromagnetic complexes with easy-axis spin symmetry (where frustration effects may be even more pronounced), ferromagnetic versions, or models in a magnetic field (see also discussion at the end of this chapter); (ii) a three-particle periodic diatomic Toda particle ring, where the mass ratio controls nonintegrability; (ii) an a.c.-driven sine-Gordon chain, which, using established techniques, can be mapped to a one-fermion half-filled two-band model in an oscillating electric field, i.e., a time-dependent Zener tunneling problem.

3.6 Quantum dynamics of a pulsed spin system: discrete map and classical result

We now move to another important subject of quantum dynamics. Since the analysis of quantum dynamics in systems with more than one degree of freedom, like a three-spin cluster, is far from feasible, we now employ a single-spin system subjected to a periodically pulsed magnetic field.

Quantum mechanics of classical chaos in driven nonautonomous systems constitutes a very active field of contemporary physics. Considering quantum dynamics, the coherent structure of wavefunctions is greatly affected by quantum interference, leading to suppression of anomalous diffusion features characteristic of chaos and eventually to vanishing of Kolmogorov–Sinai entropy and of other characteristic exponents. If we examine a semiclassical regime, however, new phenomena can appear, not present in either classical or quantum limits. Let us consider, for example, a quantum kicked rotator (Casati *et al.*, 1979) which corresponds to the standard map in the classical limit. Analyses of wavefunctions and associated contours (Korsch and Berry, 1981) show that full development of fine structure (e.g., tendrils) on all scales is suppressed because of the finiteness of the Planck constant h – distinguishing the situation from, e.g., diffusion processes in classical fluid dynamics. This fact, however, suggests at the same time the possibility of finding tendrils with much finer structure, i.e., 'fractal' tendrils, if h is decreased.

On the other hand, predicted chaos in driven spin systems (Nakamura *et al.*, 1982; Ohta and Nakamura, 1983) has recently been realized, providing an interesting example of nonlinear dynamics in the microscopic world (see chapter 4). Most experiments to date, however, have been limited to dissipative spin-wave dynamics. Fully nonlinear dynamics for noninteracting spins, both classical and quantum, is also an attractive candidate by which to study classical chaos and its quantum counterpart. As emphasized throughout sections 3.1–3.5, the advantage of spin systems is that, owing to the finite-dimensional Hilbert space, we can readily tune the value of h without artificial truncation procedures in quantum-mechanical treatments. Further, in the experiment of spin echoes in electron spin resonance, for instance, an assembly of spin-$\frac{1}{2}$ systems behaves coherently and effectively constitutes a single large quantum spin.

In the remaining sections 3.6–3.9, we consider wavefunctions for a periodically kicked spin system and demonstrate their 'quasifractal' structures in semiclassically large-spin regions. Let us consider the single-spin Hamiltonian, common to both classical and quantum spin variables **S** (Nakamura *et al.*, 1986; Frahm and Mikeska, 1986; Haake *et al.*, 1987),

$$H = A(S_z)^2 - \mu B S_x \sum_{n=-\infty}^{\infty} \delta(t - 2\pi n), \tag{3.36}$$

where A (>0) and B (>0) are an easy-plane anisotropy and magnetic field along the x axis, respectively. The effect of dissipation is omitted in the present treatment.

Before proceeding to the quantum-mechanical treatment, we shall present brief results for classical dynamics. Here **S** is a three-component vector $\mathbf{S} = S_x\mathbf{e}_x + S_y\mathbf{e}_y + S_z\mathbf{e}_z$ (**e** is a unit vector), which obeys the equation of motion

$$\mathrm{d}\mathbf{S}/\mathrm{d}t = \mathbf{S} \times (-\delta H/\delta \mathbf{S}) = \mathbf{S} \times \left[-2AS_z\mathbf{e}_z + \mu B\mathbf{e}_x \sum_{n=-\infty}^{\infty} \delta(t - 2\pi n) \right]. \tag{3.37}$$

The magnitude $\mathbf{S}^2$ is conserved in (3.37), and is now normalized to unity. We then describe $\mathbf{S}$ in polar coordinates, i.e.,

$$\mathbf{S} = (S_x, S_y, S_z) = (\sin\theta\cos\phi, \sin\theta\sin\phi, \cos\theta). \qquad (3.38)$$

The discrete map can be constructed for successive values $\{\mathbf{S}_n\}$, where $\mathbf{S}_n$ is the value of $\mathbf{S}$ at $t = 2\pi n + 0$, i.e., just after the nth pulse, see fig. 3.7. In the interval $2\pi n + 0 \le t \le 2\pi(n+1) - 0$, the magnetic field is not operative in (3.37) and thus $\mathbf{S}_n$ is transformed into $\mathbf{\Gamma} = R_z(\alpha)\mathbf{S}_n$ at $t = 2\pi(n+1) - 0$, where the operator $R_z(\alpha)$ denotes a rotation by angle $\alpha = 4\pi A S_{nz}$ around the z axis. Then, the $(n+1)$th pulse during $2\pi(n+1) - 0 \le t \le 2\pi(n+1) + 0$ rotates $\mathbf{\Gamma}$ by angle $\beta = -\mu B$ around the x axis, yielding $\mathbf{S}_{n+1} = R_x(\beta)\mathbf{\Gamma}$. Eventually, the combined map $\mathbf{S}_{n+1} = R_x(\beta)R_z(\alpha)\mathbf{S}_n$ is obtained, which is written more explicitly as

$$\begin{aligned} S_{n+1,x} &= \Gamma_x, \\ S_{n+1,y} &= \Gamma_y \cos(\mu B) + \Gamma_z \sin(\mu B), \\ S_{n+1,z} &= -\Gamma_y \sin(\mu B) + \Gamma_z \cos(\mu B), \end{aligned} \qquad (3.39a)$$

with

$$\begin{aligned} \Gamma_x &= S_{nx}\cos(4\pi A S_{nz}) - S_{ny}\sin(4\pi A S_{nz}), \\ \Gamma_y &= S_{nx}\sin(4\pi A S_{nz}) + S_{ny}\cos(4\pi A S_{nz}), \\ \Gamma_z &= S_{nz}. \end{aligned} \qquad (3.39b)$$

The map in (3.39), conserving $\mathbf{S}^2$, is essentially two-dimensional. In a specific limit $\mu B \ll 1 \ll A$, $\mathbf{S}$ is almost confined to the basal plane $(\theta \sim \pi/2)$: $S_{nx} \sim \cos\phi_n$; $S_{nz} \sim \dot{\phi}_n$. Then, noting that $4\pi A S_{nz} \ll \phi_n$, our combined map is reduced to the standard map (Chirikov, 1979; Greene *et al.*, 1981) for action $\tilde{I}_n = 2AS_{nz}$ and angle $\tilde{\phi}_n = \phi_n/2\pi$ with a stochasticity parameter $k = 4\pi A\mu B$. Without taking the standard-map limit above,

Fig. 3.7 Pulse trains and successive spin values.

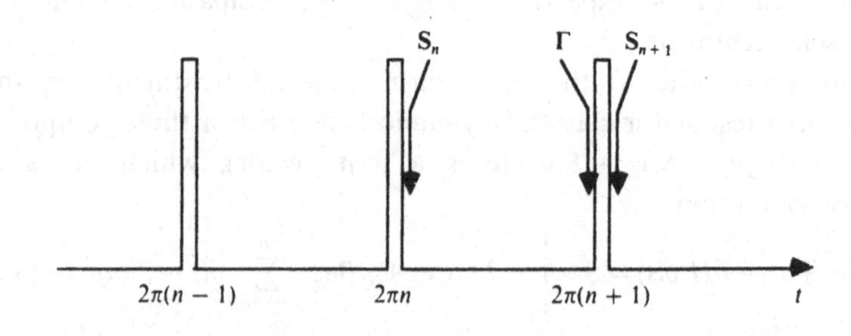

however, we here numerically iterate (3.39) with $A = 1.0$ and $0.0 \le \mu B \le 1.0$. The results are given in polar coordinates in figs. 3.8(*a*)–3.8(*c*). We see that Kolmogorov–Arnold–Moser (KAM) tori, which dominate in weak magnetic field regions (fig. 3.8(*a*)), begin to collapse with increasing field. Investigation of our extensive data as a function of μB indicates the

Fig. 3.8 Classical orbits in ϕ–θ plane ($0 \le \theta \le \pi$, $-\pi \le \phi \le \pi$); 41 orbits with initial values $\phi = 0.2$ and $\theta = j\pi/40$ ($j = 0, 1, \ldots, 40$) are shown: (*a*) $\mu B = 0.01$; (*b*) $\mu B = 0.20$; (*c*) $\mu B = 1.0$.

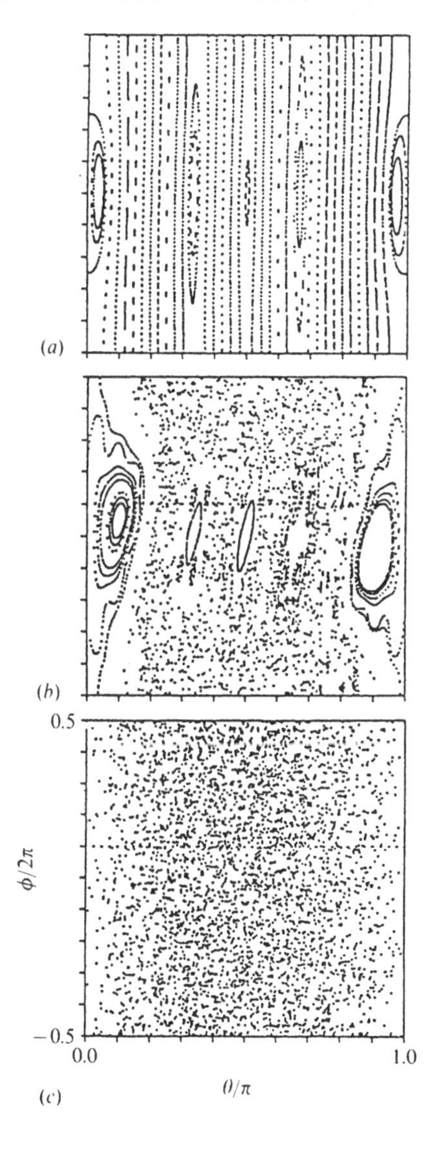

presence of two characteristic fields, $\mu B_1 \simeq 0.1$ and $\mu B_2 \simeq 0.5$, where the fraction of chaotic trajectories increases strongly and the last KAM torus disappears, respectively.

What are the quantum analogs of KAM tori and of chaos exhibited above? What is the quantum-mechanical indicator of the transition to chaos? In the next sections, we embark upon the investigation of these queries.

3.7 Quantum-mechanical treatment

In this section, we outline a quantum-mechanical framework of the corresponding quantum dynamics. The wavefunction Ψ is governed by the time-dependent Schrödinger equation

$$i\hbar \, \mathrm{d}\Psi/\mathrm{d}t = H\Psi, \tag{3.40}$$

where H is the quantum version of (3.36). We solve this equation after rewriting it in matrix form. By using basis kets $\{|m\rangle\}$ defined by $S_z|m\rangle = \hbar m|m\rangle$ ($m = -S, -S+1, \ldots, S$), Ψ is written as

$$\Psi = \sum_{m=-S}^{S} C_m(t)|m\rangle \tag{3.41}$$

and we have a coefficient vector $\mathbf{C}$ with mth component $C_m(t)$. $\mathbf{C}$ obeys the matrix equation

$$i\hbar \, \mathrm{d}\mathbf{C}/\mathrm{d}t = \tilde{H}\mathbf{C}. \tag{3.42}$$

$\tilde{H}$ is a $(2S+1) \times (2S+1)$ real-symmetric matrix:

$$\tilde{H} \equiv \tilde{H}_0 + \tilde{V} \sum_{n=-\infty}^{\infty} \delta(t - 2\pi n) \tag{3.43}$$

with

$$(\tilde{H}_0)_{m,m'} = A\hbar^2 m^2 \delta_{m,m'} \tag{3.44}$$

and

$$(\tilde{V})_{m,m'} = -(\mu B/2)\hbar\{[\hat{S}^2 - m'(m'+1)]^{1/2}\delta_{m,m'+1} + [\hat{S}^2 - m'(m'-1)]^{1/2}\delta_{m,m'-1}\}, \tag{3.45}$$

where $\hat{S} = [S(S+1)]^{1/2}$. Noting the periodicity $\tilde{H}(t+2\pi) = \tilde{H}(t)$, we apply Floquet's theorem. The solution of (3.42) just after the mth pulse is

$$\mathbf{C}(2\pi n + 0) = \sum_{\alpha} [\exp(-2\pi i n E_\alpha/\hbar)][\mathbf{X}_\alpha^\dagger \cdot \mathbf{C}(+0)]\mathbf{X}_\alpha, \tag{3.46}$$

where $\{E_\alpha\}$ and $\{X_\alpha\}$ are the quasienergies and eigenstate vectors, respectively, obtained by solving the eigenvalue problem

$$\tilde{U}\mathbf{X}_\alpha = \exp(-2\pi i E_\alpha/\hbar)\mathbf{X}_\alpha. \tag{3.47}$$

$\mathbf{X}_\alpha^\dagger$ is the Hermitian conjugate of $\mathbf{X}_\alpha$. Here U is a one-cycle unitary matrix defined in terms of the time-ordering operator T as follows:

$$\begin{aligned}\tilde{U} &= T\exp\left[\int_{+0}^{2\pi+0}(-i/\hbar)\tilde{H}(t')\,dt'\right] \\ &= \exp[(-i/\hbar)\tilde{V}]\exp[(-i/\hbar)2\pi\tilde{H}_0].\end{aligned} \tag{3.48}$$

To make the numerical diagonalization of $\tilde{U}$ more tractable, we note the following points. First, because of the symmetry of (3.36) with respect to the reversal of the quantization axis ($S_z \to -S_z$), both (3.42) and (3.48) can be decomposed into decoupled even- and odd-parity parts. Second, we take $\hbar = \hat{S}^{-1}$, i.e., $\mathbf{S}^2 = \hbar^2\hat{S}^2 = 1$ so that the observable magnitude $\mathbf{S}^2$ maintains its classical value. We thereby have $\hbar$ irrational for any integer value of S, ensuring the pseudo-randomness of matrix elements in (3.48).

We choose to describe quasienergy states

$$|\alpha\rangle = \sum_{m=-S}^{S}(\mathbf{X}_\alpha)_m|m\rangle \tag{3.49}$$

in terms of spin-coherent states, i.e., minimum-uncertainty states (Radcliffe, 1971; Klauder and Skagerstam, 1985)

$$|\theta, \phi\rangle = \exp[-i\theta(S_x\sin\phi - S_y\cos\phi)]|-S\rangle \tag{3.50}$$

and

$$\langle\theta, \phi|\alpha\rangle = \sum_{m=-S}^{S}(\mathbf{X}_\alpha)_m\langle\theta, \phi|m\rangle. \tag{3.51}$$

The probability density function $P_\alpha(\theta, \phi) = \mathcal{N}|\langle\theta, \phi|\alpha\rangle|^2$ mimics classical orbits well. The factor $\mathcal{N}$ $(=(2S+1)/4\pi)$ is due to normalization over the surface of a sphere of unit radius. In the case of the state $|\pi/2, 0\rangle$ (i.e., minimum uncertainty packet around $\theta = \pi/2$, $\phi = 0$), for example, this probability density gives

$$\mathcal{N}^{-1}P(\theta, \phi) = |\langle\theta, \phi|\pi/2, 0\rangle|^2 = \cos^{4S}(\lambda/2) \tag{3.52}$$

with $\cos\lambda = \sin\theta\cos\phi$. For $S \gg 1$,

$$\begin{aligned}\mathcal{N}^{-1}P(\theta, \phi) &= [(1+\cos\lambda)/2]^{2S} \simeq (1-(\theta-\pi/2)^2/4-\phi^2/4)^{2S} \\ &\simeq \exp[-S(\theta-\pi/2)^2/2]\exp(-S\phi^2/2),\end{aligned} \tag{3.53}$$

which indicates a minimum linear scale of $O(\hat{S}^{-1/2})$ in the (θ, ϕ) plane.

3.8 Quasi-eigenstates and fractals

We choose $S = 128$ and diagonalize (3.48) in each of even- and odd-parity manifolds (Nakamura *et al.*, 1986). Figures 3.9(*a*)–3.9(*d*) show mountainous profiles of $P_\alpha(\theta, \phi)$ for various quasienergy states. Figures 3.9(*a*)

Fig. 3.9 Three-dimensional pictures of $P_\alpha(\theta, \phi)$ for $S = 128$. (This spin magnitude is retained throughout in figs. 3.9–3.16.) (*a*) $\mu B = 0.01$, $E/\hbar = 0.1439$; (*b*) $\mu B = 0.20$, $E/\hbar = 0.3849$; (*c*) $\mu B = 0.20$, $E/\hbar = 0.1296$; (*d*) $\mu B = 1.0$, $E/\hbar = 0.9006$.

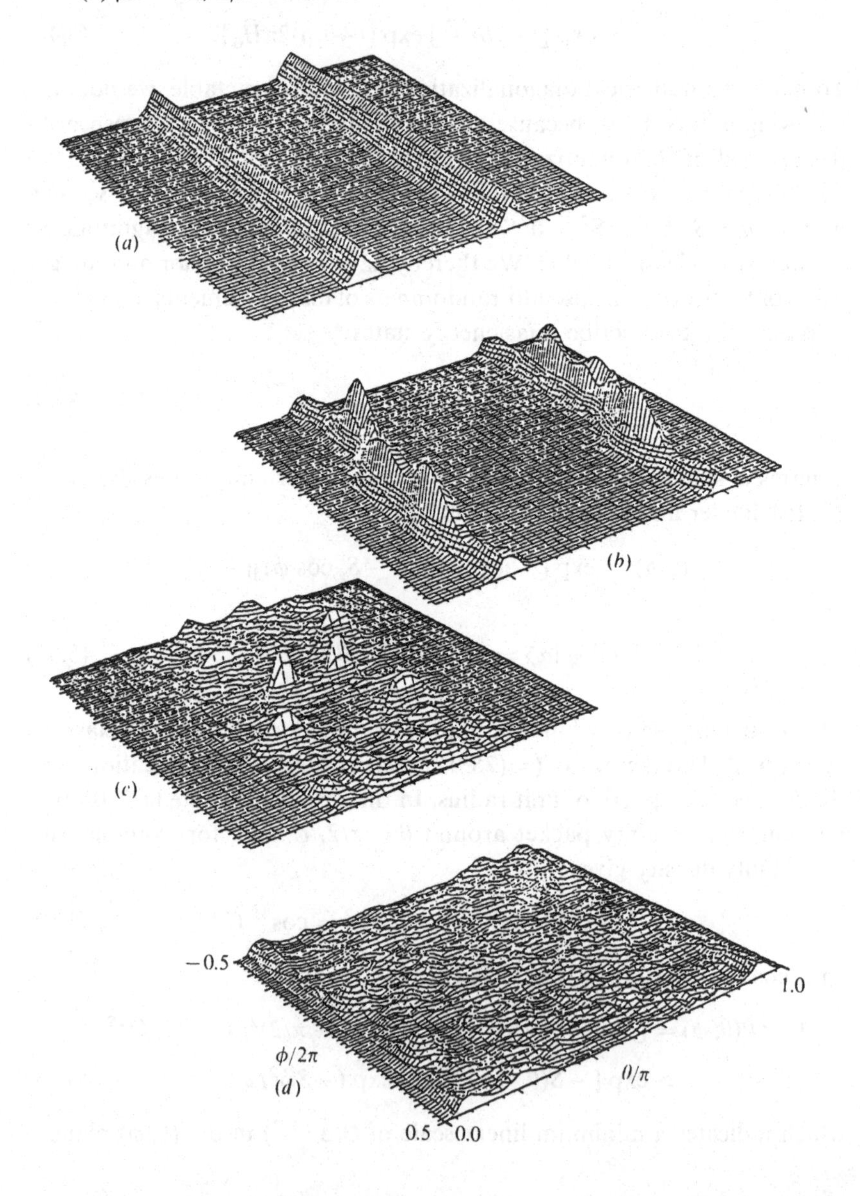

and 3.9(d) correspond typically to a classical KAM torus for $\mu B \leq \mu B_1$ and global chaos for $\mu B \geq \mu B_2$, respectively (compare figs. 3.8(a) and 3.8(c)). Figures 3.9(b) and 3.9(c) correspond to a deformed KAM torus and bounded chaos, respectively, in the region $\mu B_1 \leq \mu B \leq \mu B_2$ (compare fig. 3.8(b)). These results indicate closer examination of wavefunction localization and structure. The contour map of $P_\alpha(\theta, \phi)$ is given in fig. 3.10(a),

Fig. 3.10 Contour map for $P_\alpha(\theta, \phi)$ of fig. 3.9(d): (a) Five contour levels with height equal to $0.004l$ ($l = 1, 2, \ldots, 5$) (left-hand panel); single contour ($l = 1$) level (right-hand panel). (b) Scale-dependent binary patterns $(P_\alpha)_\varepsilon$ (from left, $\varepsilon = \varepsilon_0 \to 2\varepsilon_0 \to 2^2\varepsilon_0$). Black meshes are used for $+1$ phase. The height is $h_c = 0.004$, i.e., $\nu = 0.0818$. (This h_c value is used throughout in figs. 3.10(b), 3.11 and 3.12.)

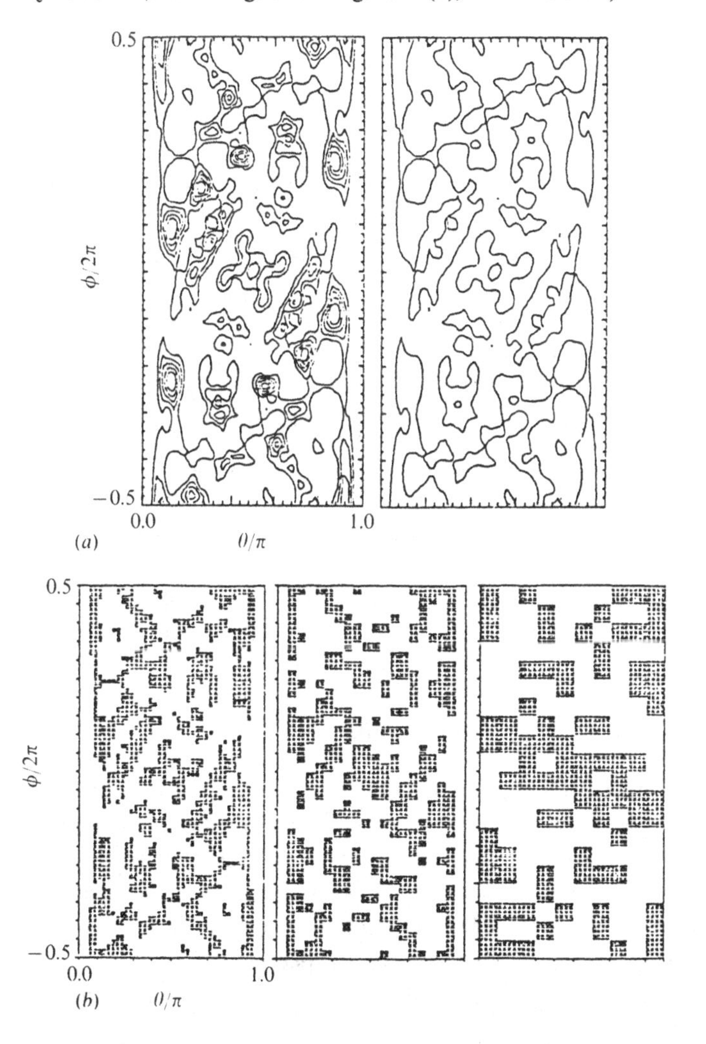

where several contours are depicted for the chaotic pattern in fig. 3.9(*d*). We clearly see tendrils or whorls with very fine structures, similar to fractal objects. We should remark that in the case of a quantum kicked rotator (see chapter 1), localization of wavefunctions is typical in a fully chaotic region because the localization length l is smaller than the linear size of momentum space (i.e., dimension d of Hilbert space). In contrast, in the present spin system $l \sim \hbar^{-2} \sim S^2$ and $d \sim S$. For $S \gg 1$, therefore, $l \gg d$ and localization phenomena seldom occur.

In an attempt to quantify $P_\alpha(\theta, \phi)$ in terms of fractals (Mandelbrot, 1982), we define scale-dependent binary patterns by coarse graining of a projection of the three-dimensional structures $P_\alpha(\theta, \phi)$. For a linear scale ε, we consider the square $\varepsilon \times \varepsilon$ mesh $A(\varepsilon)$ around the position (θ, ϕ) and assign $+1$ to it if the condition

$$\int_{(\theta,\phi) \subset A(\varepsilon)} P_\alpha(\theta, \phi)\, \mathrm{d}\Omega \Big/ \int_{(\theta,\phi) \subset A(\varepsilon)} \mathrm{d}\Omega \geq h_c \tag{3.54}$$

is satisfied, and -1 otherwise, where $\mathrm{d}\Omega = \sin\theta\, \mathrm{d}\theta\, \mathrm{d}\phi$. For all meshes in the (θ, ϕ) plane, this procedure yields the two-dimensional binary phase pattern $(P_\alpha)_\varepsilon$ on a resolution ε. The left-hand side of the above inequality is the average of P_α over surface elements of a sphere of unit radius and $h_c \equiv 4\pi v/(2S + 1)$ is a scale-independent height with a convenient fractional factor v. The value of ε increases as $\varepsilon = 2^{n-1}\varepsilon_0$ $(n = 1, 2, \ldots)$ with ε_0 the minimum scale of $O(h^{1/2}) = O(S^{-1/2})$. Both the nodal and vertical structures of $P_\alpha(\theta, \phi)$ are naturally described by $(P_\alpha)_\varepsilon$. Figure 3.10(*b*) shows variation of binary-phase patterns with increasing scales $\varepsilon = \varepsilon_0 \to 2\varepsilon_0 \to 2^2\varepsilon_0$ with $\varepsilon_0 = \pi/40$, appropriate for $S = 128$. The most striking fractal feature appears in the drastic variation of perimeter lengths for phase-pattern boundaries as the grid scale ε is changed.

In fig. 3.11, ε dependences of perimeter lengths L (Lebesgue measures) are shown for regular, fig. 3.9(*a*), and irregular, fig. 3.9(*d*), wavefunctions. They fit the scaling law

$$-\mathrm{d}(\ln L)/\mathrm{d}(\ln \varepsilon) = D_f, \tag{3.55}$$

where D_f is a fractal dimension. We find $D_f = 1.0$ and 1.58 for wavefunctions related to KAM torus and chaos, respectively. We have calculated D_f for all even-parity quasienergy states. In fig. 3.12, the distribution of D_f is shown for several values of μB. The average of D_f over quasienergies, $\langle D_f \rangle$, shows plateaux for $\mu B \leq 0.1$ and $\mu B \geq 0.5$, with $\langle D_f \rangle \simeq 1.14$ and 1.62 in the former and latter regions, respectively. For $0.1 \leq \mu B \leq 0.5$, $\langle D_f \rangle$ increases with μB. Large fluctuations of D_f in this intermediate region reflect the coexistence of KAM tori and chaos in the classical limit. The critical values $\mu B \simeq 0.1$ and 0.5 are in close agreement with μB_1 and μB_2, respectively, observed in the classical treatment. This finding strongly suggests that, in contrast to quantum kicked rotators, the persistence of quantum diffusion, rather than its suppression, seems to be characteristic of the present spin system for large S. When the height h_c is varied by

Fig. 3.11 The ε dependence of perimeter length on logarithmic scales: triangles and circles correspond to figs. 3.9(*a*) and 3.9(*d*), respectively.

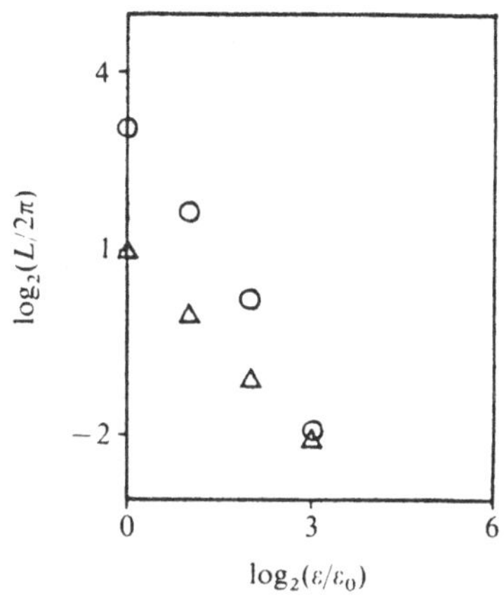

Fig. 3.12 Field dependence of D_f distributions. The triangle-in-square symbol indicates the average value $\langle D_f \rangle$. D_f values are calculated by least-squares fits to data of fig. 3.11.

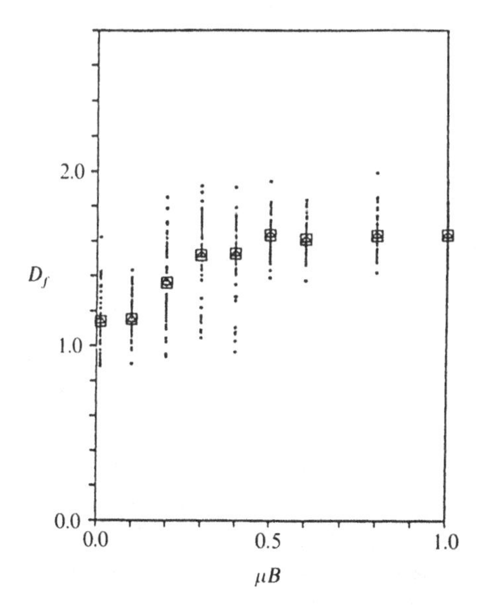

$\pm 20\%$, values D_f show little change in classically regular regimes but definite changes in chaotic regimes. Nevertheless, the general trend described above is found to be unaltered.

In this way, the spin-coherent state representation of wavefunctions in a pulsed spin system shows quasifractal patterns in semiclassically large-spin regions, whose scale-dependent contours have perimeters characterized by a fractal dimension D_f. Variation of D_f with increasing magnetic field clearly coincides with the transition from predominantly KAM tori to global chaos in the classical dynamics.

Despite several distinctive features of quasi-eigenstates described above, much more important features of quantum dynamics can be understood by investigation of time evolution of wavefunctions. We proceed to analyze this problem.

3.9 Anomalous diffusion and multifractals

Using quasienergies and quasi-eigenfunctions as inputs in (3.46), we obtain the wavefunctions Ψ in (3.41) just after the nth pulse (Nakamura *et al.*, 1989). Its probability density function is given in terms of SU(2) coherent state representation as

$$P_n(\theta, \phi) = \mathcal{N} |\langle \theta, \phi | 2\pi n + 0 \rangle|^2 \tag{3.56}$$

with $\mathcal{N}$ defined below (3.51). Since fractal dimensions D_f evaluated on the basis of binary-phase patterns for $P_\alpha(\theta, \phi)$ are sensitive to the choice of h_c in the chaotic regime, it is desirable to use a more advantageous method to quantify wavefunction patterns. In this section, we characterize $P_n(\theta, \phi)$ in (3.56) in terms of the singularity spectra $f(\alpha)$ (Halsey *et al.*, 1986) which proved very useful in quantifying multifractal aspects of 'classical' chaos.

In fig. 3.13, very early stages ($n = 1, 2, 3$) of the temporal evolution of initially ($n = 0$) localized wave packets are shown. For a weak pulse ($\mu\tilde{B} \equiv \mu B/A = 0.01$), $P_n(\theta, \phi)$ shows a simple unidirectional diffusion, see figs. 3.13(*a*)–3.13(*c*), corresponding to regular behavior in classical dynamics. (Recall the result of classical dynamics in section 3.6 which indicates the presence of two characteristic fields $\mu\tilde{B}_1 \simeq 0.1$ and $\mu\tilde{B}_2 \simeq 0.5$, where the fraction of chaotic trajectories increases greatly and the last KAM torus disappears, respectively.) However, for a strong pulse ($\mu\tilde{B} = 1.0$) remarkably isotropic and irregular diffusion processes begin after the period of 'classical' stretching- and folding-type diffusion. Let us take t_c as the cross-over time at which the classical and quantum correspondence breaks down. (In the case of hyperbolic fixed points of the corresponding classical motion, $t_c \sim$ (Lyapunov exponent)$^{-1} \times \ln(\hbar^{-1})$ as shown in (1.61) (see, e.g., Berry and Balazs, 1979).) Figure 3.13 indicates $t_c = O(1)$ for the $\mu\tilde{B} = 0.01$ and 1.0 cases. The above results resemble a quantized version of abstract dynamical systems (e.g., C or K systems), in which wavefunctions have been reported to exhibit highly irregular patterns after stretching and folding (Balazs and Voros, 1989).

We now proceed to examine $P_n(\theta, \phi)$ in large-n regions ($n = 70, 90, 110$) (see fig. 3.14). While exact classical–quantum correspondence has been lost in this time region, these figures clearly maintain some images of the underlying classical dynamics. Figures 3.14(a)–3.14(c), 3.14(a')–3.14(c') and 3.14(a'')–3.14(c'') retain signatures of nonergodicity at $\mu\tilde{B} = 0.01$ and of partial ergodicity at $\mu\tilde{B} = 1.0$, respectively. In fact, localized regular structures with large amplitudes keep a quasiperiodic oscillation for

Fig. 3.13 Contour map for very early stages of $P_n(\theta, \phi)$: (o) Initial ($n = 0$) wave packet; (a)–(c) time evolution for $\mu\tilde{B} = 0.01$; (a')–(c') time evolution for $\mu\tilde{B} = 1.0$. From the left, $n = 1$, 2 and 3.

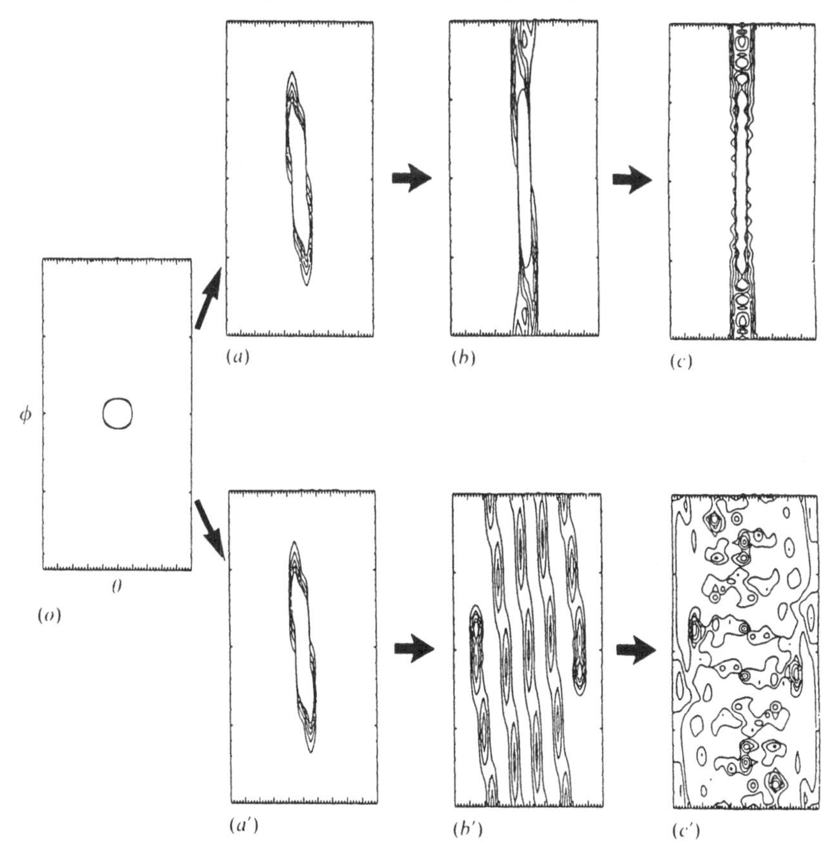

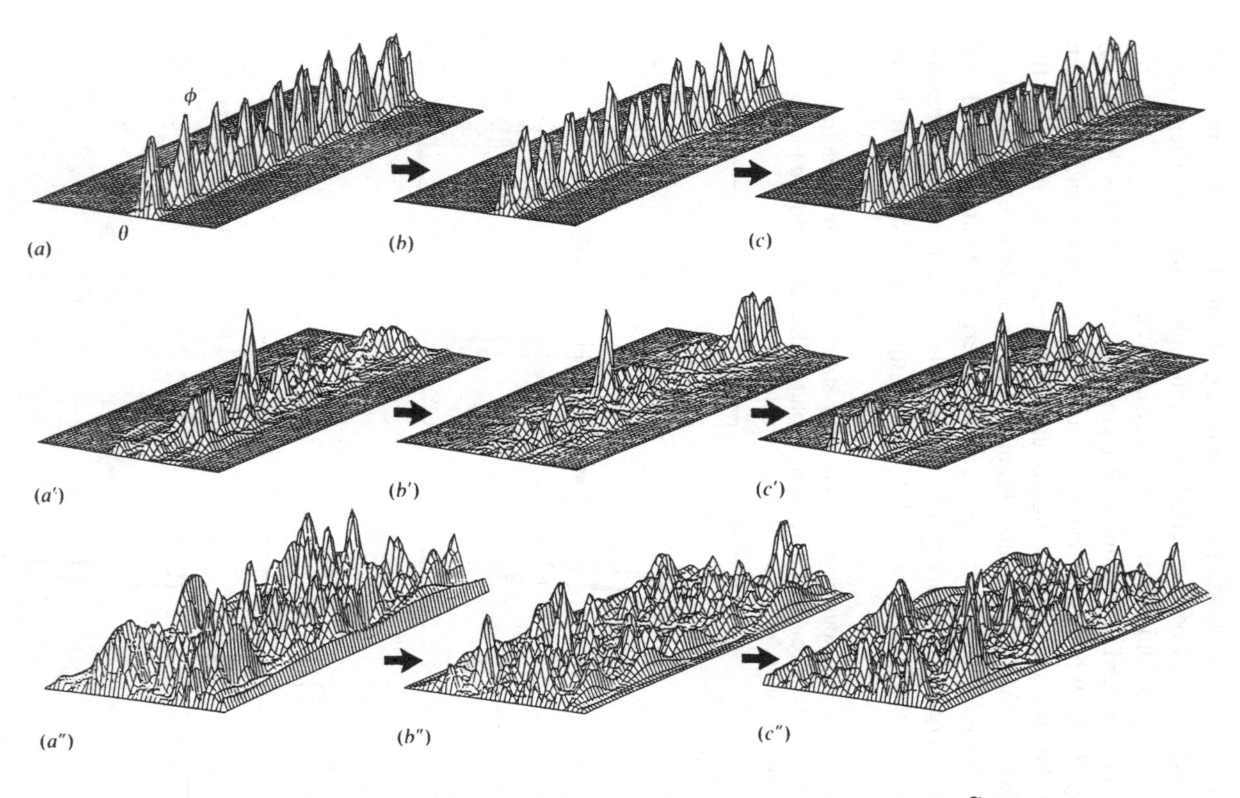

Fig. 3.14 Time evolution of three-dimensional profiles of $P_n(\theta, \phi)$ for $n \gg 1$: (a)–(c) $\mu\tilde{B} = 0.01$; (a′)–(c′) $\mu\tilde{B} = 0.2$; (a″)–(c″) $\mu\tilde{B} = 1.0$. From the left, $n = 70$, 90 and 110.

$\mu\tilde{B} = 0.01$ and fine structures with small amplitudes continue to occupy the global phase space for $\mu\tilde{B} = 1.0$. For $\mu\tilde{B} = 0.2$, fine structures continue to occupy a limited portion of phase space.

We now try to quantify $P_n(\theta, \phi)$ in terms of multifractals (Halsey *et al.*, 1986; Pietronero and Siebesma, 1986; Gutzwiller and Mandelbrot, 1988). The multifractal is a convenient tool in understanding modern nonlinear dynamics. Since it is used often in this book, details will be explained below. Any given object with a self-similar structure can be characterized by a fractal dimension, but, in general, this dimension does not have a unique value, showing fluctuations in phase space. The multifractal or $f(\alpha)$ spectrum is an important criterion characterizing the fluctuations above. Let us divide a phase space into boxes of individual linear size l. Then, the probability P_i for a given global object to occupy the ith box is given by

$$P_i \sim l^{\alpha}, \tag{3.57}$$

where the singularity α may be interpreted as a local fractal dimension. On the other hand, the number for which α falls in the range $\alpha' \leq \alpha \leq \alpha' + \mathrm{d}\alpha'$ is given by

$$N(\alpha')\,\mathrm{d}\alpha' \sim \mathrm{d}\alpha'\,\rho(\alpha')l^{-f(\alpha')}. \tag{3.58}$$

In (3.58), $f(\alpha')$ represents a fractal dimension for a subset with singularity α'.

To actual fractal objects, we can assign generalized dimensions D_q given by

$$D_q = \lim_{l \to 0} \frac{1}{q-1}(\ln \Gamma(q))/\ln l \tag{3.59a}$$

with the 'partition function'

$$\Gamma(q) = \sum_i P_i^q. \tag{3.59b}$$

Note that D_0, D_1 and D_2 denote the capacity, information and correlation dimensions, respectively. Using (3.57)–(3.58) in (3.59*b*), we have

$$\Gamma(q) = \int \mathrm{d}\alpha'\,\rho(\alpha')l^{-f(\alpha')}l^{q\alpha'}$$
$$\sim \rho(\alpha(q))l^{\tau(q)} \tag{3.60}$$

with $\tau(q) \equiv q\alpha(q) - f(\alpha(q))$, where a saddle-point method has been used in the integration (3.60). D_q thereby becomes

$$D_q = \tau(q)/(q-1). \tag{3.61}$$

Noting the saddle-point conditions in (3.60) ($\mathrm{d}f(\alpha(q))/\mathrm{d}\alpha = q$ with $\mathrm{d}^2f(\alpha(q))/\mathrm{d}^2\alpha < 0$), we have the $f(\alpha)$ spectrum expressed by the Legendre transformation, i.e., via parametric relation between α and f as

$$\alpha = \mathrm{d}\tau/\mathrm{d}q, \tag{3.62a}$$

$$f(\alpha) = q \cdot \mathrm{d}\tau/\mathrm{d}q - \tau(q). \tag{3.62b}$$

From a practical standpoint, we first evaluate $\Gamma(q)$ in (3.59*b*) as a function of l and then obtain D_q in (3.59*a*) so as to reach $\tau(q)$ in (3.61). The $f(\alpha)$ spectrum in (3.62) eventually means that subsets with fractal dimension α have a distribution characterized by fractal dimension f. As recognized in (3.59), features of $f(\alpha)$ in $q < 0$ and $q > 0$ regions characterize subsets with small and large probabilities, respectively. We note several obvious results: (i) $f(\alpha) = \alpha$ at $q = 1$; (ii) $f(\alpha) = D_0$ (the fractal dimension) at $q = 0$; (iii) $\alpha(\pm\infty) = D_{\pm\infty}$, thanks to the finiteness of $f(\alpha(q))$. From the last result, α is found to distribute in the range $D_\infty \leq \alpha \leq D_{-\infty}$.

Since $P_n(\theta, \phi)$ is already normalized to unity in the θ–ϕ plane, the calculation of the singularity spectra $f(\alpha)$ is straightforward: for a linear scale l, we consider the square $l \times l$ mesh $A_i(l)$ around the position (θ, ϕ) and calculate

$$P_{n,i}(l) = \int_{(\theta,\phi) \subset A_i(l)} P_n(\theta, \phi) \sin\theta \, \mathrm{d}\theta \, \mathrm{d}\phi. \tag{3.63}$$

Summing $P^q_{n,i}(l)$ over all meshes, we obtain the partition function $\Gamma(q, l)$. The scaling property of $\Gamma(q, l)$ is then examined by changing l according to $l = l_0 \times 2^m$ ($m = 0, 1, 2, \ldots$) with $l_0 = O(S^{-1/2})$. The scaling exponents τ_q thus obtained are used to find $f(\alpha)$. It should be noted, however, that our numerical data $P_n(\theta, \phi)$ are reliable only to the order of 10^{-5}. Using them as inputs, we can obtain wide scaling regions for $\Gamma(q, l)$ in the case $q \geq 0$, but it is difficult to explore sufficiently wide scaling regions in the case $q < 0$. So our analyses of $f(\alpha)$ below will be limited to the $q \geq 0$ case. This restriction does not prevent us studying general tendencies of fluctuations of singularities or local dimension α. Figure 3.15 represents $f(\alpha)$ with $q \geq 0$ for several $\mu\tilde{B}$ values at a fixed time $n = 90$. We find that fluctuations of α for $\mu\tilde{B} = 0.01$ and 1.0 fall into a narrow range and those for $\mu\tilde{B} = 0.2$ extend over a much wider range. The large fluctuation in the latter signifies inhomogeneous distribution of measure $P_n(\theta, \phi)$ in fig. 3.14(b'), which reflects the coexistence of classical KAM orbits and localized chaos in a transitional region leading to global chaos. This large fluctuation is reminiscent of critical fluctuations at an equilibrium phase transition. The relatively small fluctuation for $\mu\tilde{B} = 1.0$ signifies uniform distribution of measures in figs. 3.14(b''). Note that the enhanced fluctuations described above will not be observed in quantized K or C systems (Balazs and Voros, 1989) whose classical versions are homogeneously unstable and have neither KAM tori nor a transitional region.

Using our data for $f(\alpha)$ with $q \geq 0$, we now estimate the effective range of fluctuations $\alpha^*_{\rm min} \leq \alpha \leq \alpha^*_{\rm max}$, where $\alpha^*_{\rm max}$ and $\alpha^*_{\rm min}$ denote the values at which $f(\alpha)$ is a maximum (i.e., fractal dimension) and two-thirds its maximum (an arbitrary choice). For $\mu\tilde{B} = 0.2$ at $n = 90$, for example, $\alpha^*_{\rm max} = 1.98 \pm 0.02$ and $\alpha^*_{\rm min} = 1.35 \pm 0.02$. (The error bars apply for $n < 130$.) In fig. 3.16, the time dependences of the effective ranges thus introduced are shown for the interval $30 \leq n \leq 130$ in steps of 20. The features in fig. 3.15, which have now been quantified, are found to persist throughout temporal evolution. Careful examinations indicate: (1) the effective range of α shows distinctive temporal variations for $\mu\tilde{B} = 0.2$; (2) on the other hand, it remains almost unchanged for $\mu\tilde{B} = 1.0$ (despite the absence of dissipation in the present system), which reflects a well-organized ergodicity in this case.

The anomalous diffusion feature of semiclassical wavefunctions is thus summarized as follows. Despite the complete absence of classical and quantum correspondence, long-time behavior of semiclassical wavefunctions maintains ergodic and nonergodic features possessed by

Fig. 3.15 Plots of $f(\alpha)$ in regime $q \geq 0$ at $n = 90$. Squares, circles and triangles correspond to $\mu\tilde{B} = 0.01$, 0.2 and 1.0, respectively.

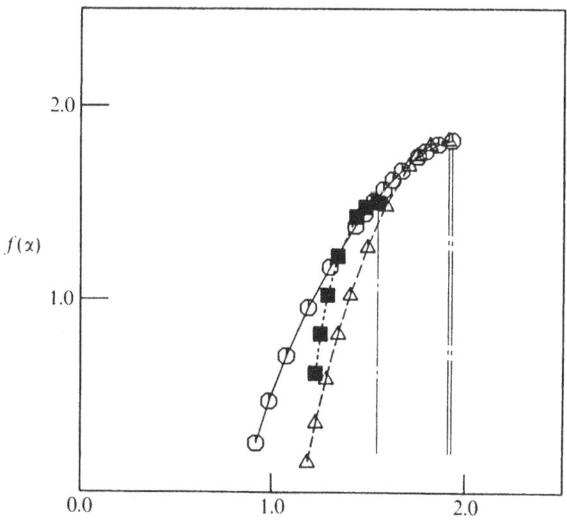

the underlying classical dynamics; enhanced fluctuation of their local dimensions in a transitional region leading to global chaos persists throughout time evolution, which is reminiscent of critical fluctuations at an equilibrium phase transition.

Mixing and ergodic features of classical chaos have helped to establish relationships with the formalism of equilibrium statistical mechanics (Bowen, 1975; Ruelle, 1978). In the field of quantum chaos, most literal definitions of classical chaos lose their significance. Nonetheless, complicated behaviors in the quantum mechanical treatment of chaotic systems are still found, as shown in this study. It seems that the characterization given here is a vehicle for more profound understanding of these complexities.

To mention the experimental relevance, quantum spin dynamics in magnetic fields has always raised challenging problems throughout the history of quantum mechanics. We can address, for example, Stern–Gerlach's discovery of quantized spins in the early days of quantum theory and recent experimental verification of fiber-bundle structures (i.e., double-valuedness of wavefunctions) of neutron spins. At present, we possess an intriguing possibility of testing the idea of quantum chaos in spin dynamics. In particular, our pulsed spin system, with several modifications

Fig. 3.16 Time evolution of effective range of $f(\alpha)$. Dotted, solid and broken lines correspond to $\mu\tilde{B} = 0.01$, 0.2 and 1.0, respectively. Symbols (■, ○, △) in each range denote the fractal dimensions, i.e., peak values of $f(\alpha)$. (The range in the case of $\mu\tilde{B} = 0.01$ at $n = 110$ is suppressed because of an accidental narrowing of scaling regions which makes it difficult to obtain reliable $f(\alpha)$ values.)

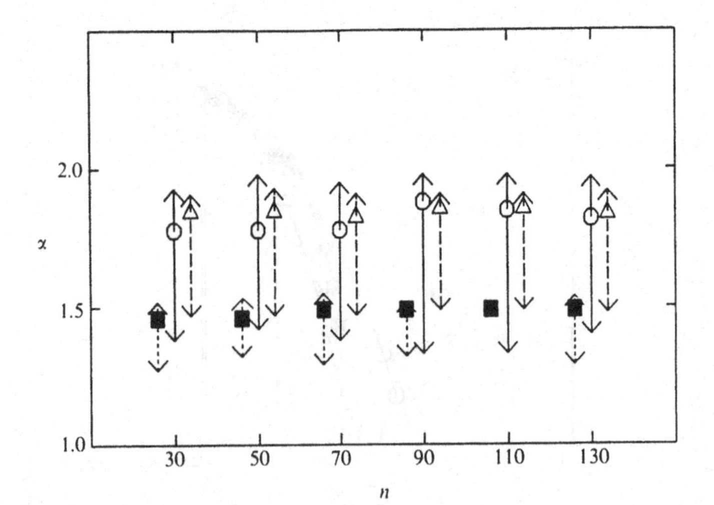

such as incorporation of dissipations, yields remarkable phenomena such as spin-echo turbulence, i.e., chaotic trains of spin-echo bursts.

The results of this chapter also open many avenues for similar studies in other fields of condensed-matter physics.

For instance, recent developments in resonating valence bond (RVB) theory (Anderson, 1987) of high-T_c superconductors have drawn wide attention to the infinite triangular lattice Heisenberg ($\sigma = 0$) antiferromagnet in two dimensions. Its elementary plaquette is precisely the three-spin cluster studied in this chapter. For the classical Heisenberg model ($\sigma = 0$) with more than three spins, the number of constants of motion is less than the degrees of freedom and hence we can expect chaos in low-energy regimes (for the antiferromagnetic case). Wen *et al.* (1989) proposed a chirality operator $\chi \equiv \mathbf{S}_1 \cdot (\mathbf{S}_2 \times \mathbf{S}_3)$ for the plaquette with spins $\mathbf{S}_1$, $\mathbf{S}_2$ and $\mathbf{S}_3$. Its global nonzero expectation value implies breaking of parity and time-reversal symmetry, a characteristic feature of RVB states. For the ground state at absolute zero, the reduced chirality $\tilde{\chi} \equiv \chi/[S(S+1)]^{3/2}$ globally takes the largest magnitude $\langle \tilde{\chi} \rangle = \pm\frac{2}{3}$ in the fully quantum limit $S = \frac{1}{2}$, and vanishes (i.e., 120° structure) in the classical limit $S = \infty$. Therefore, the chaos in low-energy regions should be a classical manifestation of chiral symmetry breaking or, conversely, RVB states are a quantum-mechanical manifestation of chaos.

On the other hand, our single-spin dynamics, endowed with suitable dissipations, will be able to simulate Bloch oscillations in a.c. driven Josephson junctions (Likharev and Zorin, 1985), because Cooper-pair operators in a small junction can be represented by quantum spins. (Note that $S^{\pm}$ $(\equiv \sum_j S_j^{\pm})$ and S^z $(\equiv \sum_j S_j^z)$ correspond, respectively, to barrier-tunneling of Cooper pairs and charge difference between superconductors bridged by the junction (Thouless, 1960).) Secondary quantum effects in Bloch oscillations are candidates in the search for evidence of quantum chaos.

4
Nonlinear dynamics in spin-wave instabilities: chaos of macroscopic quanta

This chapter is devoted to an extensive review of experimental and theoretical studies on chaos in the microscopic world of driven ferro- and antiferromagnets. We begin with a brief survey of the history of parametric instabilities of spin waves. A variety of recent experimental findings on chaos and other interesting dynamical behaviors comparable to those in fluid dynamics are described. A theoretical model derived from first principles can explain most of the experimental data.

This chapter is distinct from the others in two senses: (i) while the dynamics concern quanta (i.e., magnons) in the microscopic world, we are actually looking at semiclassical or macroscopic quantum dynamics. Therefore, the *c*-number approximation for all quantum operators is well justified, with quantum fluctuations being suppressed; (ii) the model systems here is not conservative, because the effect of dissipation is incorporated phenomenologically in the equation of motion. Readers whose interests lie exclusively in conservative quantum Hamiltonian systems may glance over this chapter.

4.1 Historical problems

Spin waves or magnons are among the typical quanta (bosons) encountered in solid-state physics. They are elementary excitations in ordered antiferro- or ferromagnets below the Curie or Néel temperatures, respectively. The magnons are characterized by energy $\hbar\omega_{\mathbf{k}}$ and momentum $\hbar\mathbf{k}$. At a finite temperature (T), their population $n_{\mathbf{k}}$ obeys the Bose–Einstein equilibrium distribution. Typically, in yttrium-iron garnet (YIG) at $T = 300$ K in a static field $H_0 = 1.5$ kOe, $n_{\mathbf{k}} \sim 10^3$ mm^{-3} for $\mathbf{k} \sim 0$ magnons. By applying the microwave magnetic field h beyond threshold, however, this population is increased catastrophically up to $n_{\mathbf{k}} \sim 10^{17}$ mm^{-3} for particular modes which satisfy the parametric resonance condition.

Spin-wave instabilities were first observed by Damon (1953) and by Bloembergen and Wang (1954) as noisy anomalous absorption which abruptly set in at a certain microwave threshold power as resonance was more strongly driven. This phenomenon has been a long-standing puzzle since then. Suhl (1957) remarked 'this situation bears a certain resemblance to the turbulent state in fluid mechanics'. The phenomenon has acquired renewed attention and was pointed out as a typical example of 'broken symmetry in dissipative structures' (Stein, 1980; Anderson, 1981). Spin-wave instabilities are caused by multi-magnon scattering processes arising from magnetic dipolar and anisotropy energies of the system. We mention two important mechanisms for such instabilities.

(i) Perpendicular pumping (Suhl, 1957): in this process, the microwave magnetic field h with frequency ω_{p} (of the order of GHz) is applied perpendicular to the static field H_0. The h field first excites the uniform ($\mathbf{k} = 0$) magnon mode with $\omega_0 \simeq \omega_{\mathrm{p}}$. Spin waves with $\mathbf{k} \neq 0$, which do not couple directly with the radiation mode, can then be excited via magnon–magnon interactions. In the case of first-order Suhl processes, three-magnon interaction is operative and the $\mathbf{k} = 0$ magnon mode decays into a pair of modes $\mathbf{k}$ and $-\mathbf{k}$ which satisfy the resonance condition $\omega_{\mathbf{k}} = \omega_{-\mathbf{k}} = \omega_{\mathrm{p}}/2$. The scattering process above is depicted in fig. 4.1(a). When the pumping field exceeds a certain threshold h_{c}, i.e., when a pair of modes are fed sufficient energy to overcome dissipation, the number of these pair modes (initially for wavevector polar angle $\theta_{\mathbf{k}} = \pi/4$) grows exponentially. Other modes not satisfying the resonance condition remain at thermal equilibrium level. Since the uniform model is *apparently* driven far off-resonance (at $\omega_0 \simeq \omega_{\mathrm{p}}/2$), this phenomenon is called 'subsidiary absorption'. In second-order Suhl processes, a spin-wave pair $\mathbf{k}$, $-\mathbf{k}$ is driven by two uniform mode magnons with $\omega_{\mathbf{k}} = \omega_{-\mathbf{k}} = \omega_0 = \omega_{\mathrm{p}}$. This is caused by the four-magnon process in fig. 4.1(b), leading to 'premature saturation', i.e., suppression of growth of the main resonance line ($\omega_0 = \omega_{\mathrm{p}}$).

(ii) Parallel pumping (Morgenthaler, 1960; Schlömann *et al.*, 1960): in this case, the h field is applied parallel to the static field H_0, and the radiation mode ($\mathbf{k} = 0$) directly excites two magnons of wavevectors $\mathbf{k}$ and $-\mathbf{k}$ with $\omega_{\mathbf{k}} = \omega_{-\mathbf{k}} = \omega_{\mathrm{p}}/2$. When the energy fed to magnon pairs begins to exceed dissipation, catastrophic increase of these pairs begins to occur for $\theta_{\mathbf{k}} = \pi/2$. A more intuitive understanding of this mechanism is made possible by providing a complementary classical picture of individual magnetic moments **m** (at lattice sites). **m** processes around the z axis parallel $\mathbf{H}_0$. Owing to the magnetic dipolar energy, etc., **m** describes a bent ellipse as in figs. 4.2 and 4.3. The frequency of the z component, $m_{\parallel}$, is twice that of the transverse component, $m_{\perp}$ ($\omega_{\parallel} = 2\omega_{\perp}$). So, when the h field with $\omega_p = \omega_{\parallel}$ causes instability of $m_{\parallel}$, instability of $m_{\perp}$ occurs simultaneously. This phenomenon resembles parametric instability of a pendulum with period T under modulation of string length with period $T/2$ (see fig. 4.2). In the perpendicular pumping case, on the other hand, the mechanism is not so tricky as that described above, and the h field induces a direct instability of $m_{\perp}$ (see fig. 4.3). Since, in magnetic insulators, magnetic moments or spins are exchange-coupled, they are not precessing in phase and their instability propagates in a form of spin waves. The resonance condition or energy–momentum conservation law corresponding to fig. 4.1(*a*)–(*c*) is depicted in fig. 4.4(*a*)–(*c*), respectively. We should emphasize that, in the above instabilities (i) and (ii), all other magnon modes lying outside the resonance surface remain at thermal level.

Despite his pioneering conjecture on the possibility of turbulence, Suhl's

Fig. 4.1 Spin-wave scattering processes: (*a*) perpendicular pumping (first-order Suhl process); (*b*) perpendicular pumping (second-order Suhl process); (*c*) parallel pumping.

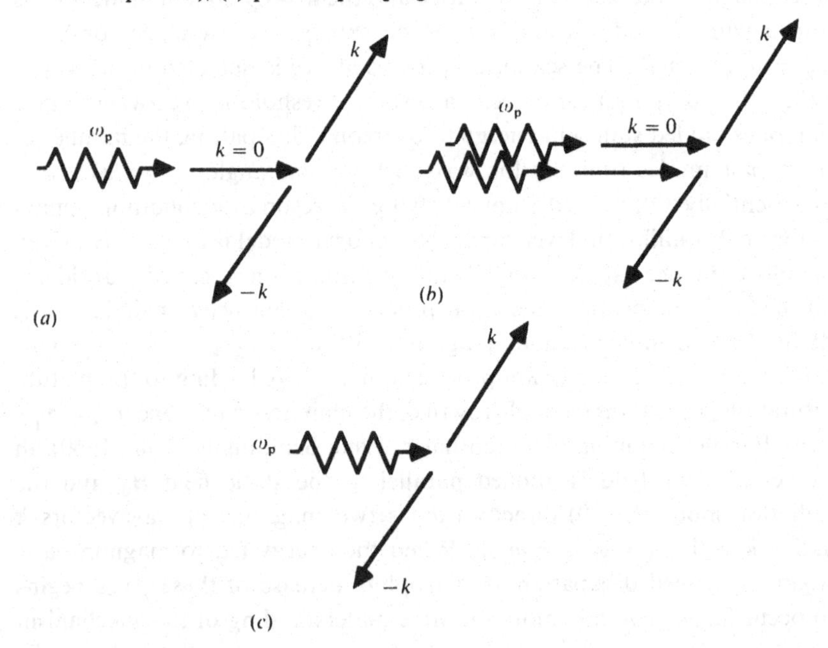

theory (Suhl, 1957) was limited to linear stability analysis of a trivial fixed point, giving no prediction of explicit dynamical behaviors in the post-threshold regime. The macroscopically-excited magnon pairs **k**, $-\mathbf{k}$ in this regime constitute standing waves in the finite sample so that the dynamical transverse component of magnetization is considered to be in a time-independent laminar state. By further increasing the microwave power, however, this quiescent state is occasionally broken by occurrence of low-frequency (10 kHz–1 MHz) periodic and nonperiodic auto-oscillations of magnon amplitudes. Hartwick *et al.* (1961) first observed these auto-oscillations (relaxation oscillations) under both parallel and perpendicular pumpings in YIG.

Zakharov *et al.* (1974) presented the S-theory which describes the nonlinear dynamics of spin waves in the post-threshold region. They proposed theoretically regular and irregular auto-oscillations. While they made a noteworthy contribution to the prehistory of modern nonlinear dynamics in the microscopic world, they showed neither strange attractors nor any route to chaos. A true renaissance of high-power magnetic resonance began in the 1980s. Nakamura *et al.* (1982) and Ohta and

Fig. 4.2 Parallel pumping: (*a*) Larmor precession of magnetic moment; (*b*) pendulum analog of (*a*).

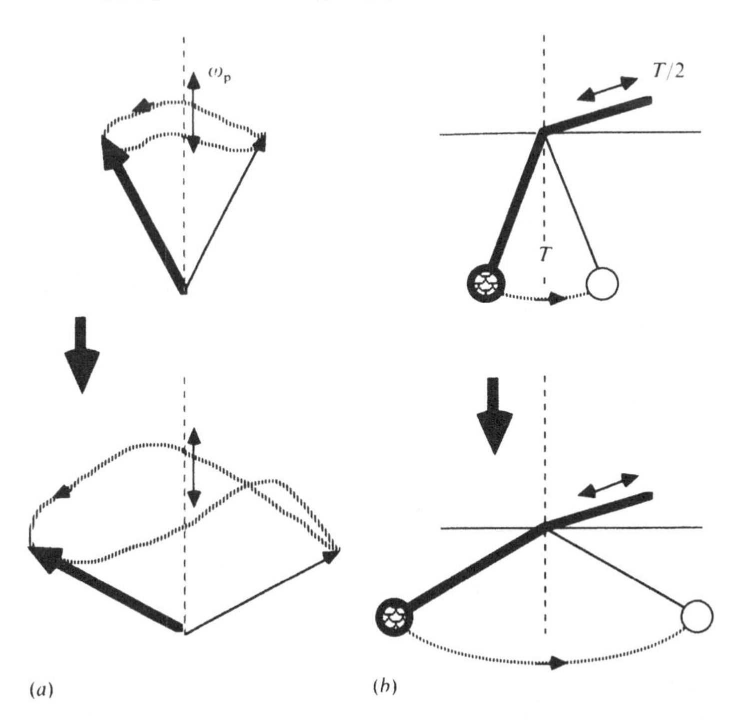

Fig. 4.3 Perpendicular pumping: (*a*) and (*b*) are the same as in fig. 4.2.

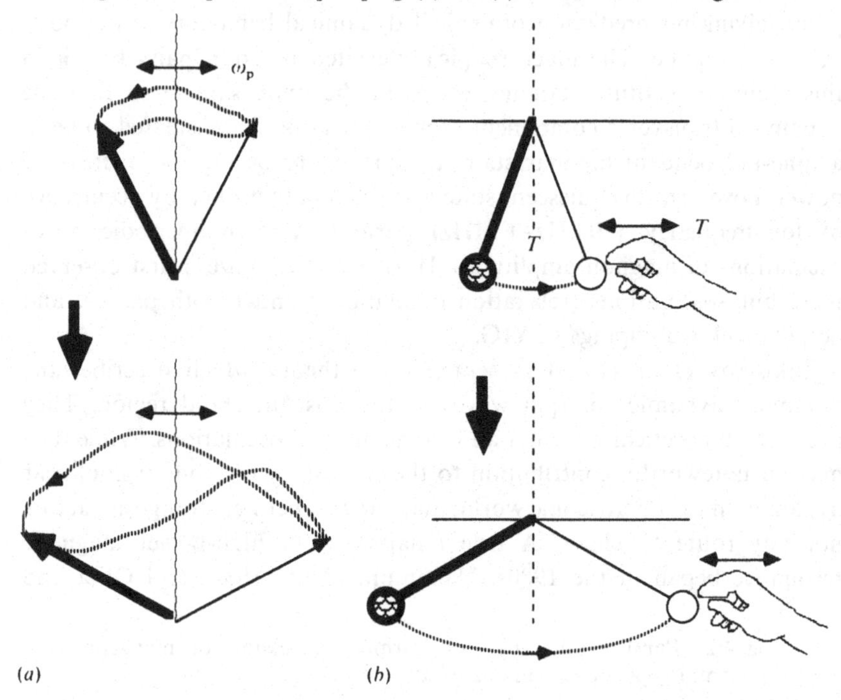

Fig. 4.4 Spin-wave dispersion curve and resonance conditions.

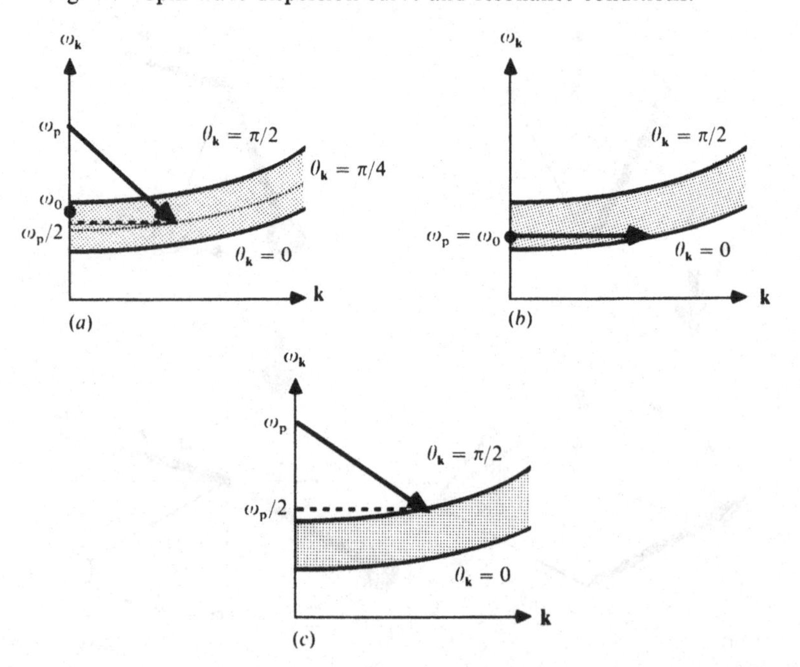

Nakamura (1983) examined numerically a truncated dynamical equation, i.e., a two-mode version of the spin-wave equation of Zakharov *et al.*, finding a period-doubling route to chaos together with the resultant strange attractor. The scaling exponents at the onset of chaos were also elucidated by them. This research has been followed by much recent experimental and theoretical work.

The great advantage of nonlinear dynamics in spin-wave systems lies in the following. (i) The theoretical background is well established. Mode identifications are possible and a few-modes dynamics is well justified, whereas in analysis (e.g., by the Lorentz model) of fluid turbulence, artificial mode-truncations are made for the Navier–Stokes equation. (ii) The characteristic time scale in the system is much less than that in fluid dynamics, so that one can promptly obtain strange attractors, Lyapunov exponents, fractal dimension, etc. (iii) Mode selections can be made by using energy–momentum conservation. Therefore, dynamics strongly depends on dispersion curves associated with magnetic structures. This fact suggests that nonlinear dynamics in the system can be a useful candidate by which to investigate material structures. (iv) There is a possibility of capturing a quantum aspect of chaos, since magnons are typical quanta in solid-state physics. In the next section, we show experimental evidence of chaos, mainly in the case of perpendicular pumping of YIG.

4.2 Experimental evidence of chaos in yttrium iron garnet (YIG)

Here we show experimental findings by Bryant *et al.* (1988a; 1988b) for first-order perpendicular pumping ('subsidiary absorption'). Readers will recognize rich structures comparable to those of chaos and turbulence in fluid dynamics. Other novel experimental results obtained by various research groups will be described separately in section 4.4..

The experiments are performed at room temperature with a sphere of pure YIG ($Y_3Fe_5O_{12}$) having a diameter $d = 0.66$ cm. YIG is a cubic insulating ferrimagnet with a Curie temperature of 559 K. The net magnetization M_s is due to the resultant of two oppositely magnetized sublattices of Fe^{3+} ions; $4\pi M_s = 1750 \pm 50$ G, and the exchange constant $\mathscr{D} = 5.4 \times 10^{-9}$ G cm^2. The Fe^{3+} ions have ground state $S = 5/2$, and consequently a weak interaction with the crystal lattice; the ferromagnet resonance linewidth is $\Delta H \simeq 0.4$ G, limited by magnon scattering at surface defects. The gyromagnetic ratio $\gamma \simeq 1.77 \times 10^7$ s^{-1} G^{-1}. The easy axis [111] was aligned parallel to the d.c. field H_0; for this orientation the resonance field is at a minimum.

Figure 4.5 indicates schematically the experimental arrangement as well as elements of the theoretical model used in the next section. Microwave power from a klystron oscillator ($P_{in} \leq 200$ mW) at frequency $f_p = \omega_p/2\pi = 9.2$ GHz is coupled via a precision attenuator, circulator and waveguide to a loop–gap resonator containing the YIG sphere and located in a uniform and stable d.c. field from a large electromagnet. The resonator is centered in the waveguide with its axis parallel to the larger transverse dimension of the waveguide. It is mounted just ahead of a sliding short circuit which can be adjusted for critical coupling to the resonator if desired. The sphere is thus subject to both a microwave magnetic field $h(t)$ and a d.c. field H_0, with $h(t) \perp H_0$ except when noted. The fields are varied in the range $0 < H_0 < 4$ kG, $0 < h < 5$ G. Microwave (pump) power $P_{in} \propto h^2$ incident on the resonator and YIG sphere is partially absorbed by damping of the resonator, uniform mode and pair of spin-wave modes, at rates Γ, γ_0 and γ_k, respectively. Power not absorbed is reflected back via the circulator to a microwave diode detector, giving a d.c. signal voltage S_0; the detector also gives an a.c. signal voltage $S(t)$ in the range 10^4–10^6 Hz, which is the sample average of the real-time signal of the collective oscillations. A power spectrum of $S(t)$ from an analog spectrum analyzer (HP model 3585A) is also recorded, with an 80 dB dynamic range, free of spurious responses.

Fig. 4.5 Schematic diagram of experimental arrangements.

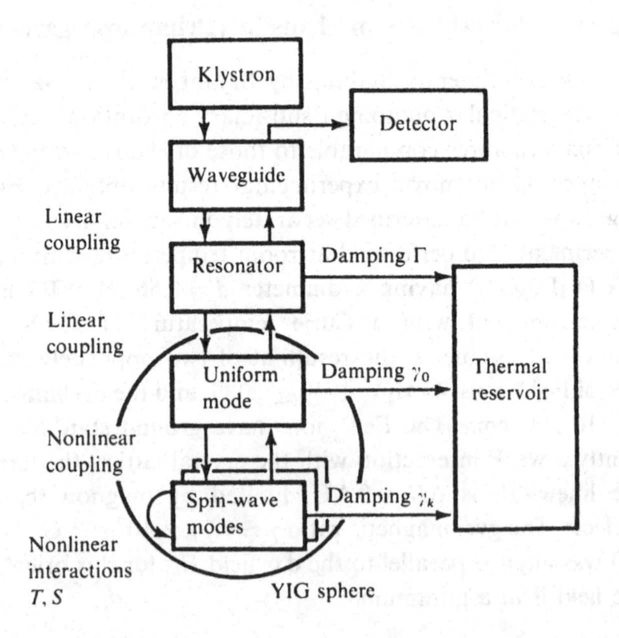

Figure 4.6 shows the phase diagram for regions and boundaries of observed behavior in parameter space (H_0, P_{in}), where P_{in} is given on a decibel scale; $P_{in} = 20 \log_{10} h$, where h is the relative microwave magnetic field; at $P_{in} = 200$ mW, $h \simeq 5$ G. In regions below the line labeled 'Suhl threshold' the system behaves linearly: the d.c. signal S_0 increases linearly with P_{in} and the a.c. signal shows only low-level wide-band detector and amplifier noise with root mean square (r.m.s.) value $S_{rms} = -70$ dB (relative level). As the Suhl threshold is crossed, either by increasing P_{in} or changing H_0 in the range $600 < H_0 < 2150$ G, there is a well-defined dip in S_0. By prebalancing the detector by an additional bridge circuit, the threshold P_{in} can be determined to within 0.05 dB. As the threshold is crossed, the a.c. signal may also change dramatically, depending on the region in parameter space. We find: (i) onset of the Suhl instability by excitation of a single spin-wave mode with very narrow linewidth (<0.5 G); (ii) when two or more modes are excited, interactions lead to auto-oscillations with a systematic dependence of frequency (10^4–10^6 Hz) on pump power, these oscillations displaying period doubling to chaos; (iii) quasiperiodicity, locking and chaos occur when three or more modes are excited; (iv) abrupt transition to wide-band power spectra (i.e., turbulence), with hysteresis; (v) irregular relaxation oscillations and

Fig. 4.6 Observed phase diagram in (H_0, P_{in}) space; $f_p = \omega_p/2\pi = 9.2$ GHz.

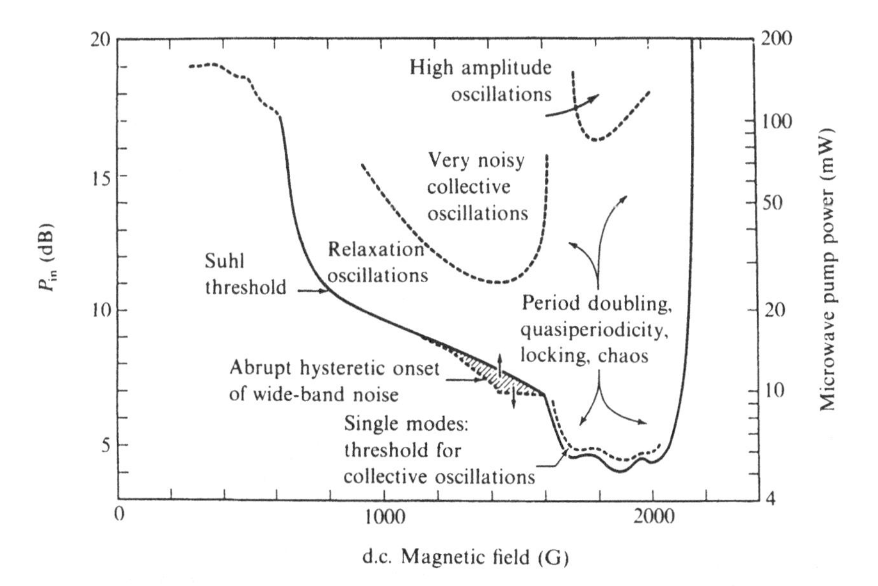

aperiodic spiking behavior; (vi) high-amplitude auto-oscillations. In the following, we describe each of the above in more detail.

(i) The Suhl threshold in the region $1600 < H_0 < 2000$ G has a rich structure, see fig. 4.7. As the field H_0 is slowly increased at constant pump power, a series of sharp dips in the d.c. signal S_0 is observed, with a spacing $\Delta H_0 = 0.156$ G, which can be understood as single modes, i.e., high-order spatial resonance modes within the sphere diameter d. For a small change in wavevector $\Delta k = \pi/d$ at constant k in (4.8) (see section 4.3.1), the computed field change is $\Delta H_0 = 2\mathscr{D}k \cdot \Delta k = 0.152$ G for $\theta_{\mathbf{k}} = 0$, using the value $k = 3 \times 10^5$ cm^{-1}. Although this elementary plane-wave calculation is in good agreement with the observed splitting, a model of spherical spin modes would be more appropriate. The individual peaks show hysteresis if the sense of the field scanning is reversed.

(ii) The first few peaks in fig. 4.7 are not accompanied by an a.c. signal $S(t)$, an indication that only single spin-wave modes are excited. However, as H_0 is increased, we arrive at a point where simultaneous excitation of two modes is possible due to mode overlap. Nonlinear mode–mode coupling may then result in onset of a low-frequency auto-oscillation signal $S(t)$ ($f_{co} \simeq 10^4$–10^6 Hz), e.g., fig. 4.8(a). In the model in section 4.3, this corresponds to a Hopf bifurcation to a limit

Fig. 4.7 Suhl threshold (partial magnification of fig. 4.6) and mode identification.

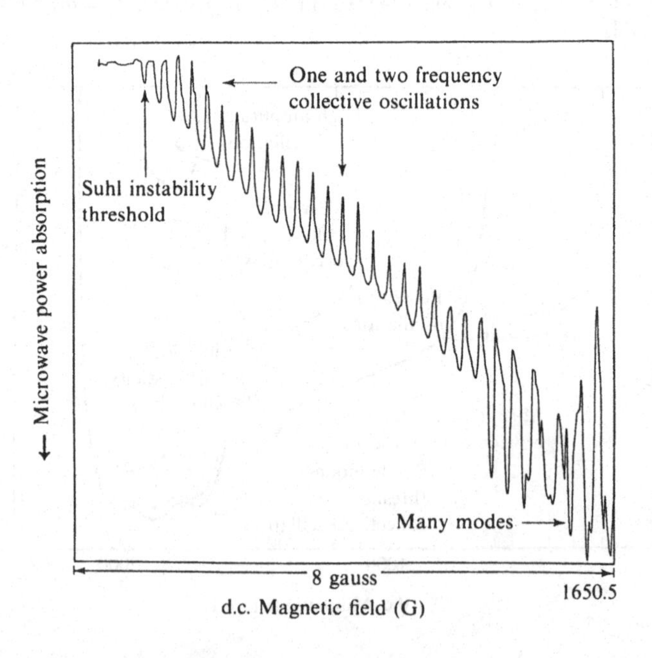

cycle when spin-wave modes are excited. These periodic collective oscillations are observed in much of the phase diagram, fig. 4.6, and display period doubling, fig. 4.8(*b*), with sharp peaks in the power spectrum, fig. 4.8(*c*), as large as 60 dB above the broad-band baseline. (Though our main concern in this section lies in the case of perpendicular pumping, similar behavior was found for parallel pumping; fig. 4.8(*d*) shows the power spectrum for an oscillation which has successively doubled to period 8.) These oscillations may show a period-doubling cascade to chaos, fig. 4.8(*e*), with very broad spectral peaks on a higher-level baseline, fig. 4.8(*f*). This cascade is induced by small variations in any of the system parameters H_0, h, f_p or crystal orientation.

(iii) As the magnetic field is further increased, the signal $S(t)$, e.g., fig. 4.9(*a*), and the power spectrum $P(f)$, e.g., fig. 4.9(*b*), give evidence for the onset of a second oscillation frequency f_2, incommensurate with the first frequency f_1. The power spectra show peaks at combination

Fig. 4.8 Observed a.c. signals are shown in (*a*), (*b*) and (*e*). Power spectra are shown in (*c*), (*d*) and (*f*).

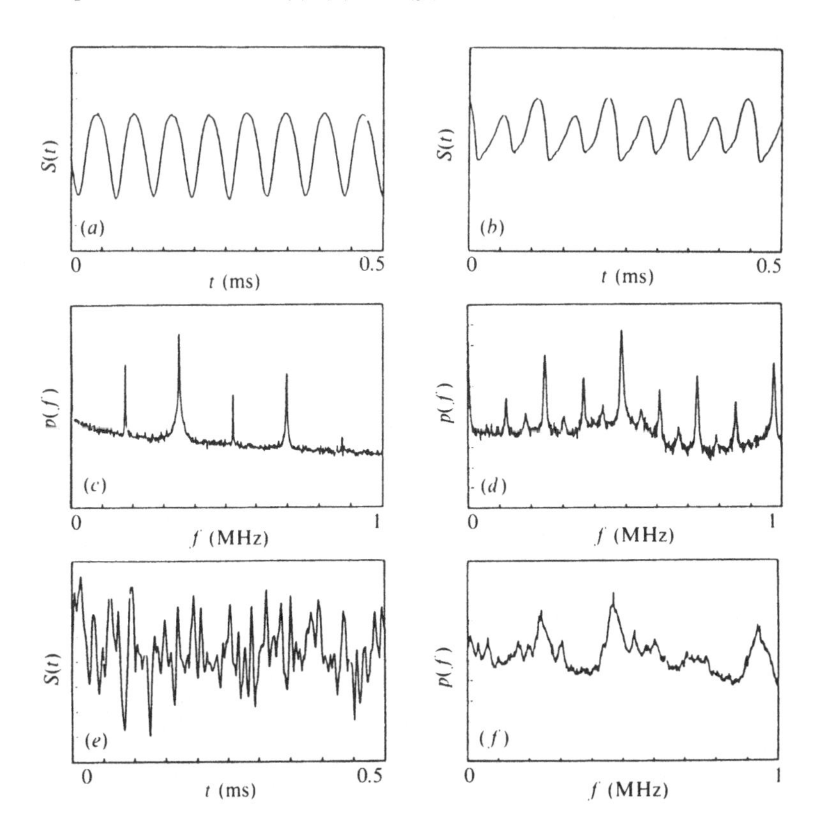

frequencies $f_{nm} = nf_1 + mf_2$, with n and m integers; intensities rapidly decay with increasing n and m. The frequencies are sensitively dependent on system parameters, and are observed to lock, $f_1/f_2 \rightarrow$ rational number, and to follow a quasiperiodic route to chaos. The real-time signal, fig. 4.9(*c*), is strikingly similar to that of two coupled pendula; fig. 4.9(*d*)

Fig. 4.9 Observed a.c. signals are shown in (*a*), (*c*), (*d*) and (*e*). Power spectra are shown in (*b*) and (*f*)–(*h*). Specifically, $\mathbf{H}_0 \parallel [100]$ in (*c*).

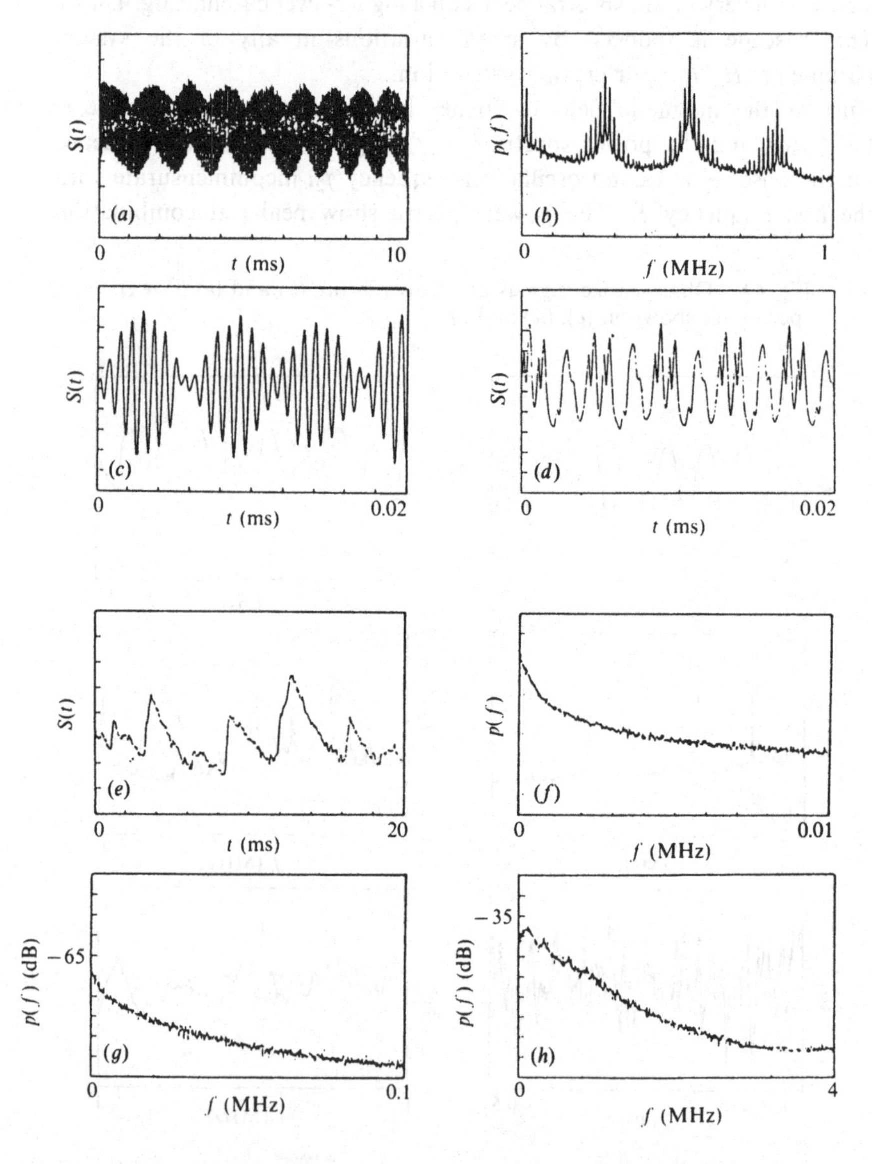

clearly shows locking at period 14. The theory in section 4.3 indicates that this quasiperiodic behavior may occur when three-spin wave modes are excited. Although we cannot clearly discern in fig. 4.7 the point at which f_2 arises, it is seen nevertheless that as H_0 is further increased, more single modes are revealed, become irregularly spaced, and overlap, a situation ripe for quasiperiodicity.

(iv) In the region $1200 \leq H_0 \leq 1600$ G in the phase diagram, fig. 4.6, as the pump power P_{in} is increased, there is an abrupt onset of wide-band (deterministic) noise ($S_{rms} \simeq -20$ dB), with no resolved spectral peaks. If P_{in} is decreased, this noise level persists until it reaches the lower level shown, then abruptly drops back to the level $S_{rms} = -70$ dB, normally observed below the Suhl threshold. Single modes as in fig. 4.7 are not observed. The hysteresis itself may possibly be understood from the model (section 4.3) as arising from subcritical symmetry-breaking bifurcations.

(v) Between 700 and 1200 G in fig. 4.6 we find just above the Suhl threshold an aperiodic signal of the general form of fig. 4.9(*e*), which has no resolved spectral peaks in the power spectrum, fig. 4.9(*f*). The peaks have a characteristic fast rise time and slow decay time known as 'relaxation' oscillations. As the pump power is increased, the times change from about 10^{-2} to about 10^{-6} s. There is a fairly abrupt transition from the power spectrum of fig. 4.9(*g*) to that of fig. 4.9(*h*). Interspersed among these wide-band high-level oscillation regions are small regions in parameter space which show periodicity, doubling and quasiperiodicity.

(vi) At high pumping power ($P_{in} > 80$ mW), with H_0 in the range 1800–2000 G, there are high-level periodic oscillations. These are typically at least an order of magnitude greater in amplitude and frequency (typical frequency 1 MHz) than those of the fine-structure regime. At these power levels, a large number of spin-wave modes become accessible, and the oscillations may be a cooperative effect involving many modes. Thus analysis of only a few modes may be of only limited applicability in this region. These oscillations exhibit all the dynamical phenomena previously described for the fine-structure regime, e.g., period doubling, quasi-periodicity and chaos. They emerge in a Hopf bifurcation at the threshold indicated in the upper right of fig. 4.6. Their emergence does not eliminate the noisy oscillations which exist below this point; however, this noise is nearly negligible compared with the oscillations when they reach full amplitude. These high-level oscillations are also observed for parallel pumping, where they occur over a much wider range of d.c. field.

The strikingly rich behaviors of nonlinear dynamics and of chaos emerging from the quantum-mechanical world of spin waves can thus be recognized. Many of these behaviors can be understood from stability analysis of fixed points and numerical iteration of the theoretical model.

4.3 Quantum theory of nonlinear spin-wave dynamics

To understand the experimental results of the previous section, we develop the quantum theory of spin-wave dynamics in the post-threshold regime. For the fundamental equation (4.30), we examine bifurcation behaviors and present numerical solutions. Theoretical results are compared with experimental data. The principal part of this section is also based on work by Bryant *et al.* (1988a; 1988b). In the S theory of Zakharov *et al.* (1974), the crucial role of resonator dynamics is suppressed and the equation of motion for spin waves in terms of 'Cooper-pair' variables is not advantageous for systematic stability analysis of fixed points. The formalism given below, on the other hand, has overcome these limitations.

4.3.1 Equation of motion for magnons in the post-threshold regime

In the present problem, we are concerned with long-wavelength magnons far from the zone boundary. It is therefore convenient to work with the magnetization field operator

$$\mathbf{M}(\mathbf{r}) = (M_x(\mathbf{r}), M_y(\mathbf{r}), M_z(\mathbf{r})),$$

rather than individual spins $\mathbf{S}_j$. This viewpoint may be called 'macroscopic quantum field theory'. We define the magnetization field

$$\mathbf{M}(\mathbf{r}) = \frac{\gamma}{2} \sum_j \boldsymbol{\sigma}_j \, \delta(\mathbf{r} - \mathbf{r}_j), \tag{4.1}$$

where $\gamma \equiv g\mu_B$ is the gyromagnetic ratio and $\boldsymbol{\sigma}$ is the Pauli matrix for the electronic spin at the site $\mathbf{r}_j$. Let us define the transverse components

$$M_\pm \equiv M_x \pm iM_y.$$

Then, $\mathbf{M}(\mathbf{r})$ is found to obey the commutation rule

$$[\mathbf{M}_+(\mathbf{r}), M_-(\mathbf{r}')] = 2\gamma M_z(\mathbf{r})\, \delta(\mathbf{r} - \mathbf{r}'). \tag{4.2}$$

In the weakly-nonlinear regime, $\mathbf{M}(\mathbf{r})$ can be transformed into bosons $a_k^\dagger, a_k$ as follows. $\mathbf{M}(\mathbf{r})$ is first transformed to bosons $a^\dagger(\mathbf{r}), a(\mathbf{r})$ by the

Holstein–Primakoff transformation and then to new bosons

$$a_{\mathbf{k}} \equiv V_s^{-1/2} \int d^3\mathbf{r}\, a(\mathbf{r})\, e^{-i\mathbf{kr}},$$

$$a_{\mathbf{k}}^{\dagger} \equiv V_s^{-1/2} \int d^3\mathbf{r}\, a^{\dagger}(\mathbf{r})\, e^{i\mathbf{kr}},$$

where V_s is the sample volume. The explicit form of the resultant transformation is

$$M_+(\mathbf{r}) = (2\gamma M_s)^{1/2}\left[1 - \frac{\gamma}{2M_s}\, a^{\dagger}(\mathbf{r})a(\mathbf{r})\right]^{1/2} a(\mathbf{r})$$

$$= M_s\left[\eta^{1/2}\sum_{\mathbf{k}} e^{-i\mathbf{kr}} a_{\mathbf{k}} - \tfrac{1}{8}\eta^{3/2}\sum_{\mathbf{k},\mathbf{k}',\mathbf{k}''} e^{i(\mathbf{k}-\mathbf{k}'-\mathbf{k}'')\mathbf{r}} a_{\mathbf{k}}^{\dagger} a_{\mathbf{k}'} a_{\mathbf{k}''} + \cdots\right], \quad (4.3a)$$

$$M_-(\mathbf{r}) = (M_+(\mathbf{r}))^{\dagger}, \quad (4.3b)$$

$$M_z = M_s - \gamma a^{\dagger}(\mathbf{r})a(\mathbf{r}) = M_s\left(1 - \frac{\eta}{2}\sum_{\mathbf{k},\mathbf{k}'} e^{i(\mathbf{k}-\mathbf{k}')\mathbf{r}} a_{\mathbf{k}}^{\dagger} a_{\mathbf{k}'} + \cdots\right), \quad (4.3c)$$

with $\eta \equiv 2\gamma/(M_s V_s)$, M_s is the saturation magnetization.

On the other hand, the Hamiltonian $\mathcal{H}$ consists of Zeeman energy $\mathcal{H}_Z$ (interaction energy with the external field) and exchange $\mathcal{H}_{ex}$, dipolar $\mathcal{H}_d$ and anistropy $\mathcal{H}_a$ energies. Each energy term is given as the volume integral of energy density expressed as a function of $\mathbf{M}(\mathbf{r})$ and its spatial derivatives (see table 4.1). By having recourse to the transformation in (4.3), however, the Hamiltonian can be written in terms of $a_{\mathbf{k}}^{\dagger}$ and $a_{\mathbf{k}}$. To diagonalize their quadratic term, we introduce the magnon operators $b_{\mathbf{k}}^{\dagger}$ and $b_{\mathbf{k}}$ by the Bogoliubov transformation:

$$a_{\mathbf{k}} = \lambda_{\mathbf{k}} b_{\mathbf{k}} - \mu_{\mathbf{k}} b_{-\mathbf{k}}^{\dagger} \quad (4.4)$$

with

$$\lambda = \left\{\frac{1}{2}\left[1 + \frac{A_{\mathbf{k}}}{(A_{\mathbf{k}}^2 - |B_{\mathbf{k}}|^2)^{1/2}}\right]\right\}^{1/2}, \quad (4.5a)$$

$$\mu = \frac{B_{\mathbf{k}}}{|B_{\mathbf{k}}|}\left\{\frac{1}{2}\left[-1 + \frac{A_{\mathbf{k}}}{(A_{\mathbf{k}}^2 - |B_{\mathbf{k}}|^2)^{1/2}}\right]\right\}^{1/2}, \quad (4.5b)$$

where $A_{\mathbf{k}}$ and $B_{\mathbf{k}}$ are defined in table 4.2 and $\lambda_{\mathbf{k}}$ and $\mu_{\mathbf{k}}$ satisfy $\lambda_{\mathbf{k}}^2 - |\mu_{\mathbf{k}}|^2 = 1$. In terms of magnon operators, the Hamiltonian is rewritten as

$$\mathcal{H} = \text{const} + \mathcal{H}_p + \mathcal{H}_2 + \mathcal{H}_3 + \mathcal{H}_4 + \cdots. \quad (4.6)$$

Table 4.1 *Various energy terms.*

J, a, v and R ($\gg a$) are exchanger constant, lattice constant, volume per spin and radius of 'Lorentz sphere' in the sample, respectively.

$\mathscr{H}_{ex}$	$\int \frac{\mathscr{D}}{2M_s} \sum_{\alpha=x,y,z} (\nabla M_\alpha)^2 \, d^3\mathbf{r}$ with $\mathscr{D} \equiv 2JSa^2/\hbar\gamma$
$\mathscr{H}_Z$	$-\int \mathbf{H}_0 \cdot \mathbf{M} \, d^3\mathbf{r}$
$\mathscr{H}_d$ (ellipsoidal samples)	$-\frac{1}{2}\int \mathbf{H}_d \cdot \mathbf{M} \, d^3\mathbf{r}$ where $H_{d,\alpha}(\mathbf{r}) = \sum_{\beta=x,y,z} \Psi_{\alpha\beta}(\mathbf{r}) M_\beta - \frac{4\pi}{3} N_\alpha M_\alpha + \frac{4\pi}{3} M_\alpha$ with $\Psi_{\alpha\beta}(\mathbf{r}) \equiv -v \sum_{\lvert\mathbf{r}_i\rvert \le R} \{\lvert\mathbf{r} - \mathbf{r}_i\rvert^{-3}[\delta_{\alpha\beta} - 3(\mathbf{r} - \mathbf{r}_i)_\alpha (\mathbf{r} - \mathbf{r}_i)_\beta \lvert\mathbf{r} - \mathbf{r}_i\rvert^{-2}]\}$ and $\sum_{\alpha=x,y,z} N_\alpha = 1$
$\mathscr{H}_a$	$\int [K_1(M_x^2 M_y^2 + M_y^2 M_z^2 + M_z^2 M_x^2)/M_s^4 + K_2 M_x^2 M_y^2 M_z^2 / M_s^6] \, d^3\mathbf{r}$

The quadratic term $\mathscr{H}_2$ is now given by

$$\mathscr{H}_2 = \sum_{\mathbf{k}} \omega_{\mathbf{k}} b_{\mathbf{k}}^\dagger b_{\mathbf{k}}, \tag{4.7}$$

where

$$\begin{aligned} \omega_{\mathbf{k}}^2 &= A_{\mathbf{k}}^2 - \lvert B_{\mathbf{k}}\rvert^2 \\ &= \left(\gamma H_0 - \frac{4\pi}{3}\gamma M_s + \gamma\mathscr{D}k^2\right)\left(\gamma H_0 - \frac{4\pi}{3}\gamma M_s + \gamma\mathscr{D}k^2 + 4\pi\gamma M_s \sin^2\theta_{\mathbf{k}}\right), \end{aligned} \tag{4.8}$$

and $\theta_{\mathbf{k}}$ is the angle between $\mathbf{k}$ and $\mathbf{H}_0$. Equation (4.8) gives the overall dispersion relation for magnons. The band structure of the dispersion curve in fig. 4.4 is caused by the $\theta_{\mathbf{k}}$ dependence in (4.8). The cubic term is

Table 4.2 *Definitions of coefficients.*

Lower part of table gives anisotropy and demagnetization factors. (Note that $k_+ = k_x + ik_y$.)

$A_{\mathbf{k}}$	$\gamma(H_0 + \mathscr{D}k^2) + 4\pi\gamma M_s\left[\delta_{\mathbf{k},0}N_{D+} - N_z + N_{A+} + (1-\delta_{\mathbf{k},0})\dfrac{\lvert k_+\rvert^2}{2k^2}\right]$		
$B_{\mathbf{k}}$	$4\pi\gamma M_s\left[\delta_{\mathbf{k},0}N_{D-} + N_{A-} + (1-\delta_{\mathbf{k},0})\dfrac{k_+^2}{2k^2}\right]$		
$L_{\mathbf{k}}$	$4\pi\gamma M_s\dfrac{k_z k_+}{2k^2}(1-\delta_{\mathbf{k},0})$		
$E_{\mathbf{k}}$	$\gamma\mathscr{D}k^2 + 4\pi\gamma M_s\left[\delta_{\mathbf{k},0}N_z + (1-\delta_{\mathbf{k},0})\dfrac{k_z^2}{k^2}\right]$		
$Q_{\mathbf{k}}$	$\gamma\mathscr{D}k^2 + 4\pi\gamma M_s\left[\delta_{\mathbf{k},0}N_{D+} + (1-\delta_{\mathbf{k},0})\dfrac{\lvert k_+\rvert^2}{2k^2} + N_{A+}\right]$		
$F_{\mathbf{k}}$	$4\pi\gamma M_s\left[(1-\delta_{\mathbf{k},0})\dfrac{k_+^2}{2k^2} + \delta_{\mathbf{k},0}N_{D-} + N_{A-}\right]$		
N_{A+}	$z \parallel [001]$ $2K_1/4\pi M_s^2$	$z \parallel [111]$ $-(\frac{4}{3}K_1 + \frac{4}{9}K_2)/4\pi M_s^2$	$z \parallel [110]$ $(-\frac{1}{2}K_1 + \frac{1}{4}K_2)/4\pi M_s^2$
N_{A-}	0	0	$(\frac{3}{2}K_1 + \frac{1}{4}K_2)/4\pi M_s^2$
$N_{D\pm}$	$(N_x \pm N_y)/2$		

$$\mathscr{H}_3 = -\eta^{1/2}\sum_{\mathbf{k},\mathbf{k}'}[W_{\mathbf{k},\mathbf{k}'}b_{\mathbf{k}}b_{\mathbf{k}'}b_{-(\mathbf{k}+\mathbf{k}')} + \text{H.c.} + V_{\mathbf{k},\mathbf{k}'}b_{\mathbf{k}}b_{\mathbf{k}'}b^{\dagger}_{\mathbf{k}+\mathbf{k}'} + \text{H.c.}], \tag{4.9}$$

where $W_{\mathbf{k},\mathbf{k}'}$ and $V_{\mathbf{k},\mathbf{k}'}$ are given in table 4.3, and H.c. denotes Hermitian conjugate. Keeping only two-modes coupling terms with slow time dependence, $\mathscr{H}_4$ is given as

$$\mathscr{H}_4 = \sum_{\mathbf{k},\mathbf{k}'}(T^{(0)}_{\mathbf{k},\mathbf{k}'}b_{\mathbf{k}}b^{\dagger}_{\mathbf{k}}b_{\mathbf{k}'}b^{\dagger}_{\mathbf{k}'} + \tfrac{1}{2}S^{(0)}_{\mathbf{k},\mathbf{k}'}b^{\dagger}_{\mathbf{k}}b^{\dagger}_{-\mathbf{k}}b_{\mathbf{k}'}b_{-\mathbf{k}'}). \tag{4.10}$$

Table 4.3 also includes expressions for $T^{(0)}_{\mathbf{k},\mathbf{k}'}$ and $S^{(0)}_{\mathbf{k},\mathbf{k}'}$. The pumping term, which comes from the volume integral $-\int \mathbf{h}(t)\cdot\mathbf{M}\,d^3\mathbf{r}$, has the resultant

Table 4.3 *Coupling constants for three- and four magnon processes.* $L_{\mathbf{k}}$, $E_{\mathbf{k}}$, $Q_{\mathbf{k}}$ and $F_{\mathbf{k}}$ are defined in table 4.2.

$W_{\mathbf{k},\mathbf{k}'}$	$L_{\mathbf{k}}\mu_{\mathbf{k}}^*\mu_{\mathbf{k}'}^*\lambda_{\mathbf{k}+\mathbf{k}'} - L_{\mathbf{k}}^*\lambda_{\mathbf{k}}\lambda_{\mathbf{k}'}\mu_{\mathbf{k}+\mathbf{k}'}^*$
$V_{\mathbf{k},\mathbf{k}'}$	$L_{\mathbf{k}}^*\lambda_{\mathbf{k}}\lambda_{\mathbf{k}'}\lambda_{\mathbf{k}+\mathbf{k}'} + L_{\mathbf{k}'}^*\mu_{\mathbf{k}}^*\lambda_{\mathbf{k}'}\mu_{\mathbf{k}+\mathbf{k}'} + L_{\mathbf{k}+\mathbf{k}'}^*\lambda_{\mathbf{k}}\mu_{\mathbf{k}'}^*\mu_{\mathbf{k}+\mathbf{k}'} - L_{\mathbf{k}}\mu_{\mathbf{k}}^*\mu_{\mathbf{k}'}^*\mu_{\mathbf{k}+\mathbf{k}'}$ $- L_{\mathbf{k}'}\lambda_{\mathbf{k}}\mu_{\mathbf{k}'}^*\lambda_{\mathbf{k}+\mathbf{k}'} - L_{\mathbf{k}+\mathbf{k}'}\mu_{\mathbf{k}}^*\lambda_{\mathbf{k}'}\lambda_{\mathbf{k}+\mathbf{k}'}$
$\frac{1}{\eta}T_{\mathbf{k},\mathbf{k}'}^{(0)}$	$\frac{1}{8}(2E_0 - Q_{\mathbf{k}} - Q_{\mathbf{k}'})(\lambda_{\mathbf{k}}^2\lambda_{\mathbf{k}'}^2 + \|\mu_{\mathbf{k}}\|^2\|\mu_{\mathbf{k}'}\|^2)$ $+ \frac{1}{2}(E_{\mathbf{k}-\mathbf{k}'} + E_{\mathbf{k}+\mathbf{k}'} - Q_{\mathbf{k}} - Q_{\mathbf{k}'})\lambda_{\mathbf{k}}\lambda_{\mathbf{k}'}\mu_{\mathbf{k}}\mu_{\mathbf{k}'}^*$ $+ \frac{1}{8}[\{(2F_{\mathbf{k}}^* + F_{\mathbf{k}'}^*)(\mu_{\mathbf{k}}\lambda_{\mathbf{k}}\lambda_{\mathbf{k}'}^2 + \lambda_{\mathbf{k}}\mu_{\mathbf{k}}\|\mu_{\mathbf{k}'}\|^2)$ $+ (F_{\mathbf{k}}^* + 2F_{\mathbf{k}'}^*)(\mu_{\mathbf{k}'}\lambda_{\mathbf{k}}^2\lambda_{\mathbf{k}'} + \lambda_{\mathbf{k}'}\|\mu_{\mathbf{k}}\|^2\mu_{\mathbf{k}'})\} + \text{'H.c.'}]$
$\frac{1}{\eta}S_{\mathbf{k},\mathbf{k}'}^{(0)}$	$\frac{1}{4}(2E_{\mathbf{k}-\mathbf{k}'} - Q_{\mathbf{k}} - Q_{\mathbf{k}'})(\lambda_{\mathbf{k}}^2\lambda_{\mathbf{k}'}^2 + \mu_{\mathbf{k}'}^2(\mu_{\mathbf{k}}^*)^2)$ $+ (E_{\mathbf{k}+\mathbf{k}'} + E_0 - Q_{\mathbf{k}} - Q_{\mathbf{k}'})\lambda_{\mathbf{k}'}\lambda_{\mathbf{k}}\mu_{\mathbf{k}}\mu_{\mathbf{k}'}^*$ $+ \frac{1}{4}[\{(F_{\mathbf{k}}^* + 2F_{\mathbf{k}'}^*)\mu_{\mathbf{k}}\lambda_{\mathbf{k}'}^2\lambda_{\mathbf{k}} + (2F_{\mathbf{k}}^* + F_{\mathbf{k}'}^*)\lambda_{\mathbf{k}'}\mu_{\mathbf{k}}^2\mu_{\mathbf{k}'}^2\} + \text{'H.c.'}]$
$\frac{1}{\eta}\Delta T_{\mathbf{k},\mathbf{k}'}$	$\frac{\omega_{\mathbf{k}+\mathbf{k}'}}{(\omega_{\mathbf{k}+\mathbf{k}'} - \omega_{\mathbf{k}} - \omega_{\mathbf{k}'})^2}V_{\mathbf{k}',\mathbf{k}}^*V_{\mathbf{k},\mathbf{k}'} + \frac{1}{\omega_0}[(V_{0,\mathbf{k}} + V_{\mathbf{k},0}) \times \text{'H.c.'}]$ $+ \frac{\omega_{\mathbf{k}+\mathbf{k}'}}{(\omega_{\mathbf{k}+\mathbf{k}'} + \omega_{\mathbf{k}} + \omega_{\mathbf{k}'})^2}[(W_{\mathbf{k}+\mathbf{k}',-\mathbf{k}'} + W_{-\mathbf{k}',\mathbf{k}+\mathbf{k}'} + W_{\mathbf{k}',\mathbf{k}}) \times \text{'H.c.'}]$
$\frac{1}{\eta}\Delta S_{\mathbf{k},\mathbf{k}'}$	$\frac{2\omega_0}{(\omega_0 - 2\omega_{\mathbf{k}})(\omega_0 - 2\omega_{\mathbf{k}'})} \times V_{\mathbf{k}',-\mathbf{k}'}V_{\mathbf{k},-\mathbf{k}}^*$ $+ \frac{2\omega_{\mathbf{k}-\mathbf{k}'}}{(\omega_{\mathbf{k}-\mathbf{k}'} + \omega_{\mathbf{k}'} - \omega_{\mathbf{k}})(\omega_{\mathbf{k}-\mathbf{k}'} + \omega_{\mathbf{k}} - \omega_{\mathbf{k}'})}$ $\times [(V_{\mathbf{k}-\mathbf{k}',\mathbf{k}'} + V_{\mathbf{k}',\mathbf{k}-\mathbf{k}'}) \times \text{'H.c.'}]$ $+ \frac{2\omega_0}{(\omega_0 + 2\omega_{\mathbf{k}})(\omega_0 + 2\omega_{\mathbf{k}'})}[(W_{0,-\mathbf{k}'} + W_{-\mathbf{k}',0} + W_{-\mathbf{k}',\mathbf{k}'}) \times \text{'H.c.'}]$

Note that 'H.c.' implies factors (terms) available by taking c.c. after the interchange $\mathbf{k} \leftrightarrow \mathbf{k}'$ in their predecessors.

form:

$$\mathcal{H}_p = \gamma h_z \sum_{\mathbf{k}} \left[\frac{A_{\mathbf{k}}}{\omega_{\mathbf{k}}} b_{\mathbf{k}}^\dagger b_{\mathbf{k}} - \frac{1}{2} \left(\frac{B_{\mathbf{k}}}{\omega_{\mathbf{k}}} b_{\mathbf{k}}^\dagger b_{-\mathbf{k}}^\dagger + \text{H.c.} \right) \right]$$
$$- \gamma \eta^{-1/2} [(\lambda_0 h_+ - \mu_0 h_-) b_0^\dagger + \text{H.c.}], \tag{4.11}$$

where $h_\pm = h_x \pm ih_y$.

While we have found terms in $\mathcal{H}_4$ which directly couple two spin-wave modes (of the same frequency), there are no such terms in $\mathcal{H}_3$. $\mathcal{H}_3$ is still needed, however, because there are second-order contributions from $\mathcal{H}_3$ which couple $\mathbf{k}$ and $\mathbf{k}'$. This occurs through off-resonance or virtual excitation of modes $\mathbf{k} + \mathbf{k}'$ and $\mathbf{k} - \mathbf{k}'$. Though these modes will not normally have the same frequency as $\mathbf{k}$ and $\mathbf{k}'$, they may nevertheless be forced into a weak response at this frequency, giving rise to a weak coupling between $\mathbf{k}$ and $\mathbf{k}'$ with magnitude of the same order as the terms in $\mathcal{H}_4$.

To deal with this problem, we employ another transformation to new variables $\tilde{b}_{\mathbf{k}}^\dagger$ which eliminates $\mathcal{H}_3$ (4.9). The transformation

$$b_{\mathbf{k}} = \tilde{b}_{\mathbf{k}} + \eta^{1/2} \sum_{\mathbf{k}'} \left[\frac{V_{\mathbf{k}',\mathbf{k}-\mathbf{k}'} \tilde{b}_{\mathbf{k}'} \tilde{b}_{\mathbf{k}-\mathbf{k}'}}{\omega_{\mathbf{k}} - \omega_{\mathbf{k}'} - \omega_{\mathbf{k}-\mathbf{k}'}} \times \frac{(V_{\mathbf{k},\mathbf{k}'}^* + V_{\mathbf{k}',\mathbf{k}}^*) \tilde{b}_{\mathbf{k}'}^\dagger \tilde{b}_{\mathbf{k}+\mathbf{k}'}}{\omega_{\mathbf{k}} + \omega_{\mathbf{k}'} - \omega_{\mathbf{k}+\mathbf{k}'}} \right.$$
$$\left. + \frac{(W_{\mathbf{k},\mathbf{k}'}^* + W_{\mathbf{k}',\mathbf{k}}^* + W_{\mathbf{k}',-(\mathbf{k}+\mathbf{k}')}^*) \tilde{b}_{\mathbf{k}'}^\dagger \tilde{b}_{-(\mathbf{k}'+\mathbf{k})}^\dagger}{\omega_{\mathbf{k}} + \omega_{\mathbf{k}'} + \omega_{\mathbf{k}'+\mathbf{k}}} \right] \tag{4.12}$$

indeed leads to $\mathcal{H}_3 = 0$. (Note that $\tilde{b}_{\mathbf{k}}^\dagger$ and $\tilde{b}_{\mathbf{k}'}$ do not satisfy the rigorous commutation relation for bosons. By adding an appropriate cubic term to (4.12), however, the approximate commutator ($[\tilde{b}_{\mathbf{k}}, \tilde{b}_{\mathbf{k}'}^\dagger] = \delta_{\mathbf{k},\mathbf{k}'}$ + (cubic terms in $\tilde{b}^\dagger, \tilde{b}$)) is available and consequently the Heisenberg equation of motion for $\{\tilde{b}_{\mathbf{k}}\}$ is correct as far as cubic terms.) The transformation (4.12) also generates additional quartic terms $\Delta T_{\mathbf{k},\mathbf{k}'}$ and $\Delta S_{\mathbf{k},\mathbf{k}'}$, see table 4.3, and as a result $T_{\mathbf{k},\mathbf{k}'}^{(0)}$ and $S_{\mathbf{k},\mathbf{k}'}^{(0)}$ are renormalized to new variables $T_{\mathbf{k},\mathbf{k}'} \equiv T_{\mathbf{k},\mathbf{k}'}^{(0)} + \Delta T_{\mathbf{k},\mathbf{k}'}$ and $S_{\mathbf{k},\mathbf{k}'} \equiv S_{\mathbf{k},\mathbf{k}'}^{(0)} + \Delta S_{\mathbf{k},\mathbf{k}'}$. We now have

$$\mathcal{H}_4 = \sum_{\mathbf{k},\mathbf{k}'} (T_{\mathbf{k},\mathbf{k}'} \tilde{b}_{\mathbf{k}} \tilde{b}_{\mathbf{k}}^\dagger \tilde{b}_{\mathbf{k}'} \tilde{b}_{\mathbf{k}'}^\dagger + \tfrac{1}{2} S_{\mathbf{k},\mathbf{k}'} \tilde{b}_{\mathbf{k}}^\dagger \tilde{b}_{-\mathbf{k}}^\dagger \tilde{b}_{\mathbf{k}'} \tilde{b}_{-\mathbf{k}'}). \tag{4.13}$$

From tables 4.2 and 4.3, $T_{\mathbf{k}',\mathbf{k}}^* = T_{\mathbf{k},\mathbf{k}'}$ and $S_{\mathbf{k}',\mathbf{k}}^* = S_{\mathbf{k},\mathbf{k}'}$ can be recognized, ensuring the Hermitian nature of (4.13). The quadratic part of the Hamiltonian remains invariant under the transformation (4.12):

$$\mathcal{H}_2 = \sum_{\mathbf{k}} \omega_{\mathbf{k}} \tilde{b}_{\mathbf{k}}^\dagger \tilde{b}_{\mathbf{k}}. \tag{4.14}$$

The important consequence of (4.12) is a change in the perpendicular pumping term in (4.11). In $b_{\mathbf{k}}$ notation, the transverse field $h_{\pm}$ only couples to the uniform mode b_0 which is off resonance. This in turn couples to a spin-wave pair, $b_{\mathbf{k}}, b_{-\mathbf{k}}$, via term in $\mathscr{H}_3$ like $V^*_{\mathbf{k},-\mathbf{k}} b^\dagger_{\mathbf{k}} b^\dagger_{-\mathbf{k}} b_0$. In the $\tilde{b}_{\mathbf{k}}$ notation, however, the H_3 term no longer exists. Instead, we now have new terms appearing in H_{p} whereby the external field couples directly to spin-wave pairs just as it does for the parallel pumping term. The essential part of H_{p} may now be expressed,

$$\mathscr{H}_{\mathrm{p}} = \sum_{\mathbf{k}} \left[-\tfrac{1}{2}\gamma h_z \frac{B_{\mathbf{k}}}{\omega_{\mathbf{k}}} - \gamma(\lambda_0 h_+ - \mu_0 h_-) \frac{V^*_{\mathbf{k},-\mathbf{k}}}{\omega_0 - 2\omega_{\mathbf{k}}} - \gamma(\lambda_0 h_- - \mu_0^* h_+) \frac{W^*_{0,\mathbf{k}} + W^*_{\mathbf{k},0} + W^*_{\mathbf{k},-\mathbf{k}}}{\omega_0 + 2\omega_{\mathbf{k}}} \right] \tilde{b}^\dagger_{\mathbf{k}} \tilde{b}^\dagger_{-\mathbf{k}} + \text{H.c.}, \tag{4.15}$$

where we have omitted the direct coupling terms $h_+\tilde{b}_0 + h_-\tilde{b}^\dagger_0$, since they have no effect on spin-wave pairs. If we restrict our attention to the case where demagnetization ($N_{\mathrm{D}-}$) and anisotropy ($N_{\mathrm{A}-}$) factors are both zero, as occurs, for example, with a spherical sample of YIG with $\mathbf{H}_0$ parallel to [111] or [100] (see table 4.2), then $\lambda_0 = 1$ and $\mu_0 = 0$ and (4.15) simplifies to

$$\mathscr{H}_{\mathrm{p}} = \sum_{\mathbf{k}} \left[-\tfrac{1}{2}\gamma h_z \frac{B_{\mathbf{k}}}{\omega_{\mathbf{k}}} - \gamma h_+ \left(\frac{L_{\mathbf{k}} \lambda^2_{\mathbf{k}} - L^*_{\mathbf{k}} \lambda_{\mathbf{k}} \mu_{\mathbf{k}}}{\omega_0 - 2\omega_{\mathbf{k}}} \right) - \gamma h_- \left(\frac{L^*_{\mathbf{k}} \mu^2_{\mathbf{k}} - L_{\mathbf{k}} \lambda_{\mathbf{k}} \mu_{\mathbf{k}}}{\omega_0 + 2\omega_{\mathbf{k}}} \right) \right] \tilde{b}^\dagger_{\mathbf{k}} \tilde{b}^\dagger_{-\mathbf{k}} + \text{H.c.} \tag{4.16}$$

The first term corresponds to parallel pumping by the z component of the microwave field h_z. The second term corresponds to perpendicular pumping by the component of the transverse field with counter-clockwise circular polarization. The third term corresponds to perpendicular pumping by the component with clockwise circular polarization. (This term is weak and may usually be ignored.) Both parallel and perpendicular pumpings can thus be treated in a unified way.

In experimental situations, the microwave pumping field is generated by a resonator (cavity mode) which surrounds the sample. Therefore, we should also mention the role of resonator dynamics and waveguide–resonator coupling. The resonator amplitude is represented by R^+, R and contributes a term $\mathscr{H}_R$ to the total Hamiltonian

$$\mathscr{H}_R = \omega_R R^\dagger R, \tag{4.17}$$

where ω_R is the resonant frequency. By equating the maximum field energy with $\mathscr{H}_R$ we find that the field h_R is given by

$$h_R = \left[\frac{2\pi\omega_R}{V_R}\right]^{1/2}(R + R^\dagger), \tag{4.18}$$

where V_R is the effective volume of the resonator. Since we wish to consider oblique pumping we assume that h_R is linearly polarized, lies in the x–z plane, and makes an angle θ_R with the z axis. Noting $h_z = h_R \cos\theta_R$, $h_\pm = h_R \sin\theta_R$, $\mathscr{H}_\mathrm{p}$ in (4.16) eventually becomes

$$\mathscr{H}_\mathrm{p} = \sum_{\mathbf{k}} \tilde{G}_{\mathbf{k}} R \tilde{b}_{\mathbf{k}}^\dagger \tilde{b}_{-\mathbf{k}}^\dagger + \mathrm{H.c.}, \tag{4.19}$$

where

$$\tilde{G}_{\mathbf{k}} = -\tfrac{1}{2}\gamma\left(\frac{2\pi\omega_R}{V_R}\right)^{1/2}\frac{B_{\mathbf{k}}}{\omega_{\mathbf{k}}}\cos\theta_R$$
$$-\gamma\left(\frac{2\pi\omega_R}{V_R}\right)^{1/2}\sin\theta_R\left(\frac{L_{\mathbf{k}}\lambda_{\mathbf{k}}^2 - L_{\mathbf{k}}^*\lambda_{\mathbf{k}}\mu_{\mathbf{k}}}{\omega_0 - 2\omega_{\mathbf{k}}} + \frac{L_{\mathbf{k}}^*\mu_{\mathbf{k}}^2 - L_{\mathbf{k}}\lambda_{\mathbf{k}}\mu_{\mathbf{k}}}{\omega_0 + 2\omega_{\mathbf{k}}}\right). \tag{4.20}$$

Using the whole Hamiltonian ((4.13), (4.14) and (4.16)) $\mathscr{H} = \mathscr{H}_\mathrm{p} + \mathscr{H}_2 + \mathscr{H}_4$, we obtain the Heisenberg equation of motion for magnons $\{b_{\mathbf{k}}\}$

$$\dot{\tilde{b}}_{\mathbf{k}} = \mathrm{i}[\omega_{\mathbf{k}}\tilde{b}_{\mathbf{k}} + 2\tilde{G}_{\mathbf{k}} R \tilde{b}_{-\mathbf{k}}^\dagger + \sum_{\mathbf{k}'}(2T_{\mathbf{k},\mathbf{k}'}\tilde{b}_{\mathbf{k}}\tilde{b}_{\mathbf{k}'}\tilde{b}_{\mathbf{k}'}^\dagger + S_{\mathbf{k},\mathbf{k}'}\tilde{b}_{-\mathbf{k}}^\dagger\tilde{b}_{\mathbf{k}'}\tilde{b}_{-\mathbf{k}'})]. \tag{4.21}$$

For further analysis, however, we use the following procedures. Noting the macroscopic-quantal nature of active modes, we make the c-number approximation for both magnon and resonator operators:

$$\tilde{b}_{\mathbf{k}} \to \tilde{b}_{\mathbf{k}}^{\mathrm{cl}}, \tag{4.22a}$$

$$R \to R^{\mathrm{cl}}, \tag{4.22b}$$

where cl denotes classical; then we incorporate in the equation of motion the phenomenological damping terms. Eventually, (4.21) becomes

$$\dot{\tilde{b}}_{\mathbf{k}}^{\mathrm{cl}} + \gamma_{\mathbf{k}}\tilde{b}_{\mathbf{k}}^{\mathrm{cl}} = \mathrm{i}[\omega_{\mathbf{k}}\tilde{b}_{\mathbf{k}}^{\mathrm{cl}} + 2\tilde{G}_{\mathbf{k}} R^{\mathrm{cl}}\tilde{b}_{-\mathbf{k}}^{\mathrm{cl}*} + \sum_{\mathbf{k}'}(2T_{\mathbf{k},\mathbf{k}'}\tilde{b}_{\mathbf{k}}^{\mathrm{cl}}\tilde{b}_{\mathbf{k}'}^{\mathrm{cl}}\tilde{b}_{\mathbf{k}'}^{\mathrm{cl}*}$$
$$+ S_{\mathbf{k},\mathbf{k}'}\tilde{b}_{-\mathbf{k}}^{\mathrm{cl}*}\tilde{b}_{\mathbf{k}'}^{\mathrm{cl}}\tilde{b}_{-\mathbf{k}'}^{\mathrm{cl}})], \tag{4.23}$$

where $\gamma_{\mathbf{k}}$ is the damping constant for the magnon.

In the waveguide, there are incoming and outgoing waves h_{in} and h_{out}, respectively. Both h_{in} and h_{out} are taken to be the effective amplitudes of waves at the sample location. For simplicity, we assume

$$h_{\mathrm{in}} = P_{\mathrm{in}}^{1/2}\,\mathrm{e}^{\mathrm{i}\omega_\mathrm{p} t}, \tag{4.24}$$

where P_{in} is the input power. Two components make up h_{out}: the first is

h_{in} just reflected at the end of the waveguide; the second is emission from the resonator and is assumed to be proportional to R. Thus

$$h_{\text{out}} = h_{\text{in}} + \mathrm{i}\beta R,$$

where β is a complex constant. Ignoring weak coupling to the uniform mode and following the same procedure as given below (4.21), the resonator obeys the equation

$$\dot{R}^{\text{cl}} = -(\Gamma_{\text{res}} + \Gamma_{\text{rad}})R^{\text{cl}} + \mathrm{i}[\omega_R R^{\text{cl}} + \alpha h_{\text{in}} + \sum_{\mathbf{k}} \tilde{G}_{\mathbf{k}}^* \tilde{b}_{\mathbf{k}}^{\text{cl}} \tilde{b}_{-\mathbf{k}}^{\text{cl}}]. \quad (4.25)$$

where Γ_{res} and Γ_{rad} are resistive and radiative dampings, respectively (Γ_{res} and Γ_{rad} are both assumed real), and α is a complex coupling parameter. Conservation of energy determines the relationship between α, β and Γ_{rad},

$$\Gamma_{\text{rad}} = \tfrac{1}{2}\alpha\alpha^*\omega_R, \quad (4.26a)$$

$$\beta = \alpha^*\omega_R. \quad (4.26b)$$

For critical coupling, $h_{\text{out}} = 0$ and this implies

$$\Gamma_{\text{rad}} = \Gamma_{\text{res}} = \tfrac{1}{2}\alpha\alpha^*\omega_R \equiv \frac{\omega_R}{2Q}, \quad (4.27)$$

where Q is the quality factor of the resonator. We can set $|\alpha|$ to the critical value by adjusting a single experimental parameter, e.g., the sliding short circuit at the end of the waveguide.

Equations (4.23) and (4.25) constitute fundamental equations of motion for the system depicted in fig. 4.3. Taking into account experimental situations, we can make further simplifications as follows: first, since spin-wave modes excited in pairs form standing waves, one may assume that $\tilde{b}_{-\mathbf{k}}^{\text{cl}} = \exp(\mathrm{i}q_{\mathbf{k}})\tilde{b}_{\mathbf{k}}^{\text{cl}}$ with $q_{\mathbf{k}}$ a real phase factor; second, the slow variables $c_{\mathbf{k}}$ and $\bar{R}$ are introduced as

$$\tilde{b}_{\mathbf{k}}^{\text{cl}} = c_{\mathbf{k}}\, \mathrm{e}^{-\mathrm{i}q_{\mathbf{k}}/2}\, \mathrm{e}^{\mathrm{i}\omega_p t/2}, \quad (4.28a)$$

$$R^{\text{cl}} = \bar{R}\, \mathrm{e}^{\mathrm{i}\omega_p t}; \quad (4.28b)$$

finally, noting that $\Gamma_{\text{res}}, \Gamma_{\text{rad}} \gg \gamma_{\mathbf{k}}$ (lossy resonator), we adiabatically eliminate the resonator variable $\bar{R}$ as follows. Let us assume that $\mathrm{d}\bar{R}/\mathrm{d}t = 0$ in (4.25), and we then obtain

$$\bar{R} = \frac{\mathrm{i}\sum_{\mathbf{k}} \tilde{G}_{\mathbf{k}}^* c_{\mathbf{k}}^2}{\Gamma - \mathrm{i}\,\Delta\Omega_R} \quad (4.29)$$

with $\Gamma = \Gamma_{res} + \Gamma_{rad}$ and $\Delta\Omega_R = \omega_R - \omega_p$. By substituting (4.29) into (4.23), one obtains for $c_\mathbf{k}$

$$\dot{c}_\mathbf{k} = (-\gamma_\mathbf{k} + \mathrm{i}\,\Delta\Omega_\mathbf{k})c_\mathbf{k} - G_\mathbf{k} P_{in}^{1/2} c_\mathbf{k}^* + \mathrm{i}\sum_{\mathbf{k}'} [2T_{\mathbf{k},\mathbf{k}'}|c_{\mathbf{k}'}|^2 c_\mathbf{k} + (S_{\mathbf{k},\mathbf{k}'} + U_{\mathbf{k},\mathbf{k}'})c_{\mathbf{k}'}^2 c_\mathbf{k}^*], \tag{4.30a}$$

where

$$U_{\mathbf{k},\mathbf{k}'} = 2\mathrm{i}\tilde{G}_{\mathbf{k}'}^* \tilde{G}_\mathbf{k}/(\Gamma - \mathrm{i}\,\Delta\Omega_R) \tag{4.30b}$$

and

$$G_\mathbf{k} = 2\tilde{G}_\mathbf{k}\alpha/(\Gamma - \mathrm{i}\,\Delta\Omega_R) \tag{4.30c}$$

with $\Delta\Omega_\mathbf{k} = \omega_\mathbf{k} - \omega_p/2$.

Equation (4.30) is the principal theoretical result. It should be noted that only a limited number of magnon modes join the dynamics in (4.30): most magnons lying outside the resonance surface remain at thermal level. Within the resonance surface itself, $G_\mathbf{k}$ in (4.30) takes predominant values at specific modes $\mathbf{k}$, which further reduces the number of active modes. This kind of mode truncation is quite favorable to the modern analysis of nonlinear dynamics (Ruelle and Takens, 1971).

4.3.2 Stability analysis of fixed points

In this subsection, we examine (4.30) analytically to explore the fixed points and their stability. Let us first consider the case where only one spin wave is excited. We find that $c_\mathbf{k} = 0$ is always a fixed point – this is true regardless of how many spin-wave modes are excited to nonzero values. Its stability depends on the relative strength of the damping term $-\gamma_\mathbf{k} c_\mathbf{k}$ and forcing term $-G_\mathbf{k} P_{in}^{1/2} c_\mathbf{k}^*$. An important feature to note about (4.30) is that there is inversion symmetry; if $c_\mathbf{k}(t)$ is a solution then so is $-c_\mathbf{k}(t)$. This is also true for arbitrarily many modes: the sign of each may be changed independently without affecting the validity of the solution. For the stability analysis of the fixed point $c_\mathbf{k} = 0$ we need only consider the linear part of the equation,

$$\dot{c}_\mathbf{k} = (-\gamma_\mathbf{k} + \mathrm{i}\,\Delta\Omega_\mathbf{k})c_\mathbf{k} - G_\mathbf{k} P_{in}^{1/2} c_\mathbf{k}^*. \tag{4.31}$$

Solving for the eigenvalues, we find that $c_\mathbf{k} = 0$ is stable when $|M| > 1$ where

$$M = (-\gamma_\mathbf{k} - \mathrm{i}\,\Delta\Omega_\mathbf{k})/G_\mathbf{k}^* P_{in}^{1/2}. \tag{4.32}$$

The condition $|M| = 1$ corresponds to the 'Suhl threshold' for mode $\mathbf{k}$.

Since $|M|$ is inversely related to input power P_{in}, $|M| > 1$ is below threshold and $|M| < 1$ is above threshold. Above threshold, the origin is a saddle point, as it always retains one stable eigenvalue. Immediately below threshold, the origin is a stable node with two negative real eigenvalues. However, the eigenvalues may split into a conjugate pair below a lower threshold, which corresponds to a change from a stable node to a stable focus. This occurs for $P_{\text{in}} < \Delta\Omega_{\mathbf{k}}^2/|G_{\mathbf{k}}|^2$.

We will now consider nontrivial (or nonzero) fixed points. Since we are still considering the behavior of a single mode, the equation $\dot{c}_{\mathbf{k}} = 0$ [from (4.30)] can be put in the simple form

$$M + N|c_{\mathbf{k}}|^2 = \frac{c_{\mathbf{k}}^2}{|c_{\mathbf{k}}|^2} = \text{points on unit circle,} \tag{4.33}$$

where $N = -\mathrm{i}(2T_{\mathbf{k},\mathbf{k}} + S_{\mathbf{k},\mathbf{k}} + U_{\mathbf{k},\mathbf{k}}^*)/G_{\mathbf{k}}^* P_{\text{in}}^{1/2}$ and M is given in (4.32). This equation has a simple geometrical interpretation, as shown in fig. 4.10. We plot point M and the unit circle in the complex plane. If we are

Fig. 4.10 (*a*)–(*c*) Nontrivial fixed-point analyses. (*d*) and (*e*) Bifurcation diagrams; F and F_c denote pump power and Suhl threshold, respectively; solid and dashed lines correspond to stable and unstable fixed points, respectively. Arrows in (*e*) indicate hysteresis.

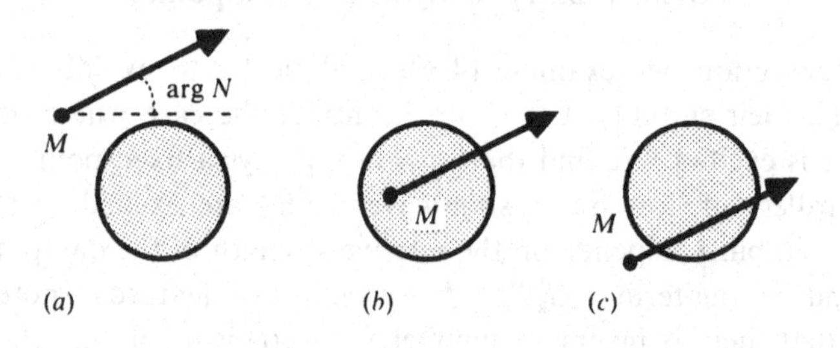

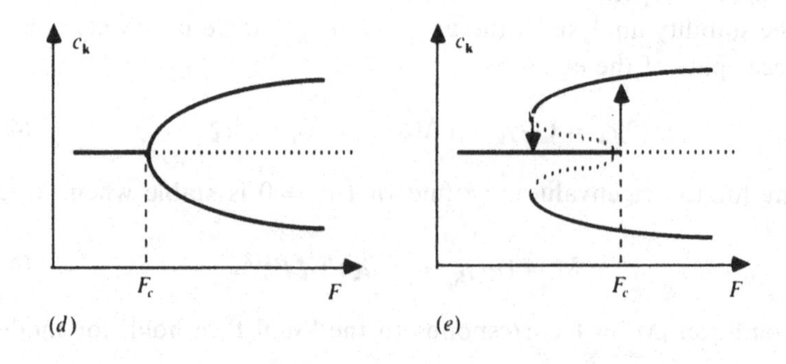

below threshold then M lies outside the circle. We draw a line from point M making an angle $\arg(N)$ with respect to the real axis. Typically this line will either miss the circle as in fig. 4.10(*a*), in which case there are no nontrivial fixed points, or it will intersect the circle at two points as in fig. 4.10(*c*), in which case there are two pairs of nontrivial fixed points $\pm c_{\mathbf{k}}^{[1]}$ and $\pm c_{\mathbf{k}}^{[2]}$. Nontrivial fixed points always come in pairs because of the symmetry of the equations mentioned previously. The transition between the two cases occurs when the line is just tangent to the circle. This is a saddle–node bifurcation with the saddle being the interaction point closest to M. Above the Suhl threshold, M is inside the circle and there is always one intersection point as in fig. 4.10(*b*). Two possibilities may occur when crossing the Suhl threshold and these are illustrated in figs. 4.10(*d*) and 4.10(*e*). In the first case, which occurs for $\mathrm{Re}\,(M/N) > 0$, we obtain a supercritical symmetry-breaking bifurcation since a complementary pair of stable nontrivial fixed points emerge from the origin as it loses stability. The second case, which occurs for $\mathrm{Re}\,(M/N) < 0$, involves existence of the saddle and node below threshold. In this case we obtain a subcritical symmetry-breaking bifurcation, where the unstable nontrivial fixed points (saddle points) converge on the origin as it changes stability. There is a hysteresis loop, as shown, because the system must jump from the origin to one of the stable nodes which are at finite amplitude.

In the event that $U_{\mathbf{k},\mathbf{k}}$ may be neglected (i.e., it is much smaller than $S_{\mathbf{k},\mathbf{k}}$ as may be the case if the resonator damping is sufficiently large, see (4.27)), the type of bifurcation can be changed by changing the sign of $\Delta\Omega_{\mathbf{k}}$. We find this experimentally in the region where fine structure is observed (see section 4.2) but here hysteresis is also a very fine effect, occurring over a very small range in parameter space. However, there is a region indicated in fig. 4.6 in which hysteresis occurs over a substantial range in parameter – much more than can be attributed to a single mode. This is probably due to a related effect in which more than one mode is simultaneously excited.

In the experiment, it was found that in certain regions aperiodic relaxation-type oscillations are observed which are characterized by alternating fast and slow phases, where the amplitude of reflected microwaves changes very rapidly in the fast and much more gradually in the slow phase, typically differing by an order of magnitude or more, see fig. 4.9(*e*). A related behavior pattern has also been observed, in which rapid spikes in the response are separated by long dormant periods of irregular length. We now present a mechanism which can explain both types of behavior and discuss some of its effects on the dynamics of the experimental system.

The simplest system which can exhibit this type of behavior is a two-mode system, represented by complex variables c_1 and c_2. Mode c_1 can be called the 'strong mode'; it is assumed that the pumping level is above the Suhl threshold for this mode. Mode c_2 can be called the 'weak mode'; it is assumed that the pumping level is below the Suhl threshold for this mode. In the absence of coupling between c_1 and c_2, we would expect that the origin of c_1 would be a saddle point so that this mode would be attracted to a nonzero fixed point, while the origin would be stable for c_2 so that this mode would decay to zero. However, due to mode–mode coupling, the stability of the origin for c_2 can be affected by the amplitude of c_1. When coupling is included, the stability criterion for $c_2 = 0$ is

$$\left|\frac{A + D|c_1|^2}{B + Fc_1^2}\right| > 1, \tag{4.34}$$

where $A = -\gamma_2 - \mathrm{i}\,\Delta\Omega_2$, $B = -G_2 P_{\mathrm{in}}^{1/2}$, $D = -2\mathrm{i}T_{2,1}$, and

$$F = -\mathrm{i}(S_{2,1} + U_{2,1}).$$

The assumption that c_2 is below its Suhl threshold (for $c_1 = 0$) implies that $|A| > |B|$. There are four general cases for the behavior of the stability of c_2 as a function of c_1. Case 1 is $|F| > |D|$. In this case, as $|c_1|$ is increased for any particular phase $\phi = \arg c_1$, a point is reached beyond which the denominator in (4.34) becomes larger in magnitude than the numerator, and stability is lost. The point at which stability is lost is a function of ϕ and has inversion symmetry as shown in fig. 4.11(*a*). Case 2 is $|F| < |D|$ and $\mathscr{K} > |BF|$, where $\mathscr{K} \equiv [(|A|^2 - |B|^2)(|D|^2 - |F|^2)]^{1/2}\ \mathrm{Re}\, AD^*$. In this case point $c_2 = 0$ is a stable fixed point for all values of c_1, as shown in fig. 4.11(*b*). Case 3 is $|F| < |D|$ and $-|BF| < \mathscr{K} < |BF|$. In this case there are two symmetrically located stability zones in the c_1 plane, as shown in fig. 4.11(*c*). Case 4 is $|F| < |D|$ and $\mathscr{K} < -|BF|$. This case has an annulus of instability as shown in fig. 4.11(*d*). The general stability boundary for all four cases can be expressed as a quadratic solution,

$$|c_1|^2 = \frac{-b \pm (b^2 - 4ac)^{1/2}}{2a}, \tag{4.35}$$

where $a = |D|^2 - |F|^2$, $b = AD^* + DA^* - BF^*\,\mathrm{e}^{-\mathrm{i}\phi} - FB^*\,\mathrm{e}^{\mathrm{i}\phi}$ and $c = |A|^2 - |B|^2$. Since for cases 2, 3 and 4, a and c are both positive, a solution for $|c_1|$ exists only if b is more negative than $-(4ac)^{1/2}$ (in which case there are two solutions).

Now that we have analyzed the stability of the weak mode, we can proceed to explain the nature of the oscillations. We will suppose that c_1

and c_2 both start at some small but finite value. Then c_1 will increase, approaching a nontrivial fixed point and c_2 will decrease towards zero. Assuming that the nontrivial fixed point for c_1 lies in (or possibly across) a zone of instability for c_2, the phase path of c_1 will eventually enter this zone. Beyond this point, c_2 will begin to increase. Under certain conditions, which we do not specify precisely here but demonstrate in section 4.3.3, this can lead to relaxation, with the fast phase occurring after c_1 reaches the instability boundary and the slow phase occurring when c_1 and c_2 both go back to values near 0. The reason that c_1 can return to a point near zero is that the origin for c_1 is a saddle point and therefore an orbit near the stable manifold may come quite close to the origin before escaping again. Orbits of the type described are nearly homoclinic since they pass very close to a saddle point in the four-dimensional c_1, c_2 space. In the event that the weak mode has a focus at the origin, the orbit may be of Silnikov type (Guckenheimer and Holmes, 1983), which is known to imply existence of horseshoes and other complicated behavior. The

Fig. 4.11 Stability curves for $c_2 = 0$ as a function of $c_1 = 0$. Wavy areas are unstable, while others are stable.

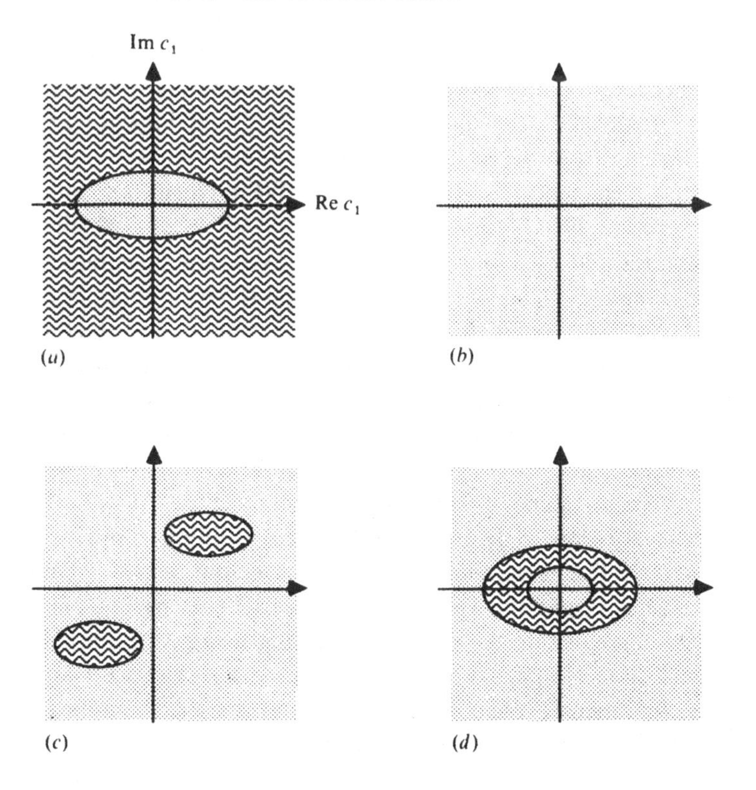

distinction between relaxation oscillations and aperiodic spiking lies in the length of the dominant phase for the weak mode. During the dormant phase, the amplitude of c_2 decays exponentially. Consequently, a moderately long dormant phase can easily result in the amplitude of c_2 decaying to thermal magnon level. This introduces stochasticity into the dynamics – something which might not ordinarily be expected for oscillators whose peak amplitude is many orders of magnitude above thermal level.

4.3.3 Numerical treatment

We proceed to explore fully the behavior of the spin-wave equation (4.30) by performing numerical analysis. Computed results are presented, particularly when two or more interacting modes are involved. They can be compared with experimental results of section 4.2.

Each spin-wave mode is represented by a complex variable $c_{\mathbf{k}}$ which contains both amplitude and phase information for that mode. From (4.30), we obtain one equation for each mode, which is coupled to all other excited modes through interaction parameters $U_{\mathbf{k},\mathbf{k}'}$, $S_{\mathbf{k},\mathbf{k}'}$ and $T_{\mathbf{k},\mathbf{k}'}$, and to the microwave pumping field through the parameter $G_{\mathbf{k}}$. Our analytic results provide rough estimates for these parameters. They cannot be specified exactly from theory for two reasons: (i) we do not know for certain which spin-wave modes in the sample are being excited and hence involved in the dynamics, and (ii) the plane-wave approximation used in section 4.2 can yield only approximate values for interaction parameters of spherical modes. We typically set spin-wave damping $\gamma_{\mathbf{k}}$ to $1 \times 10^6\ \mathrm{s}^{-1}$ and $G_{\mathbf{k}}$ to $1.414 \times 10^7\ \mathrm{W}^{-1/2}\ \mathrm{s}^{-1}$, which results in a Suhl threshold of about 5 mW as observed experimentally in the single-mode region. $T_{\mathbf{k},\mathbf{k}'}$ and $S_{\mathbf{k},\mathbf{k}'}$ are estimated to be of the order of 10^{19} or $10^{20}\ \mathrm{G}^{-2}\ \mathrm{s}^{-2}$, but may vary considerably depending on which modes are involved. In order to simulate the effect of a sequence of modes as observed experimentally, we will not assume that all modes have zero detuning (i.e., $\Delta\Omega_{\mathbf{k}} \neq 0$ in (4.30)), but will instead choose a sequence of equally spaced values for $\Delta f_{\mathbf{k}}$ which will typically extend from some negative value to some positive value, where $\Delta f_{\mathbf{k}} \equiv \Delta\Omega_{\mathbf{k}}/2\pi$. If the excitation level is low, only those modes with detunings closest to zero will be excited. The remaining modes will be below threshold and will remain at zero amplitude. From the observed field spacing of modes (0.16 G) we can estimate the frequency spacing using $\Delta f_{\mathrm{mode}} \simeq \gamma\, \Delta H/2\pi$, obtaining a value of approximately 500 kHz. As we shall see below, the essential experimental phenomena can be explained theoretically by participation of at most three spin-wave modes.

(1) Single-mode dynamics: in this case, the analytic results of section 4.3.2 determine the location and stability of all fixed points. There is always at least one stable fixed point and numerical results indicate that the system is always attracted to one of these; no periodic or chaotic attractors are observed. For appropriate parameter values, hysteresis may be observed, as indicated in the analytical treatment.

(2) Two-modes dynamics: in this case we first find periodic auto-oscillations. A particularly interesting form is observed, as shown in figs. 4.12(*a*) and 4.12(*b*) (parameters for which are given in table 4.4). Here mode 2 exhibits an asymmetric orbit while mode 1 exhibits a symmetrical orbit of twice the period. Symmetrical orbits are possible because of the inherent inversion symmetry of the equations. When asymmetric orbits occur they always come in complementary pairs ($c'(t) = -c(t)$). The nature of coupling between modes allows the type of behavior observed – since the square of c_1 appears in the equation for c_2, a sign change of c_1 (to the opposite point on the symmetric orbit) has identical influence on c_2.

By changing parameters we find that this orbit may undergo a bifurcation. There are many parameters which could be adjusted to accomplish this, such as power input or pump frequency, but in figs. 4.12(*c*) and 4.12(*d*) we have chosen to shift synchronously the detuning Δf_k of the modes. This is equivalent to shifting the value of the d.c. magnetic field in the experiment. (We assume that all modes in the sequence have identical field dependence of their frequencies.) The result is interesting: mode 1 exhibits a symmetry-breaking bifurcation while mode 2 simultaneously exhibits period doubling. Further shifting of frequencies produces a cascade of period-doubling bifurcations for both modes, leading to a chaotic orbit, see figs. 4.12(*e*) and 4.12(*f*).

The numerical study for two modes also reveals behavior similar to the relaxation oscillations and aperiodic spiking found for the experimental system. An example of this behavior is shown in fig. 4.13. The mechanism for this behavior was discussed previously in section 4.3.2. There is a 'strong mode' which is above threshold, and a 'weak mode' which is initially below threshold but which can become excited for brief periods when sufficient excitation is supplied via nonlinear coupling of the strong mode. Characteristically, there is a slow or dormant phase during which the weak mode decays, tending increasingly closer to zero, and the strong mode changes at a relatively slow rate. At a certain critical point in the orbit of the strong mode, which is indicated by an arrow in fig. 4.13(*a*), a fast or active phase commences during which both modes (figs. 4.13(*a*) and 4.13(*b*)) change rapidly. This is typically two or more orders of magnitude faster, and also shorter in duration, than the slow phase. The decay of the mode during the dormant phase may be extreme – it has been observed in some cases of numerical study to decay by over ten orders of magnitude. This will easily take any experimental system to thermal level, thus introducing a stochastic element into the dynamics. Orbits with a short dormant phase tend to have a 'relaxation oscillation'

Fig. 4.12 (*a*)–(*e*) Computed trajectories with Poincaré section marked by circles. (*f*) is the power spectrum for (*e*). In (*a*), (*b*) $\Delta f_1 = -300$ kHz, $\Delta f_2 = 200$ kHz; (*c*), (*d*) $\Delta f_1 = 385$ kHz, $\Delta f_2 = 115$ kHz; (*e*), (*f*) $\Delta f_1 = -410$ kHz, $\Delta f_2 = 90$ kHz. Other parameter values are given in table 4.4.

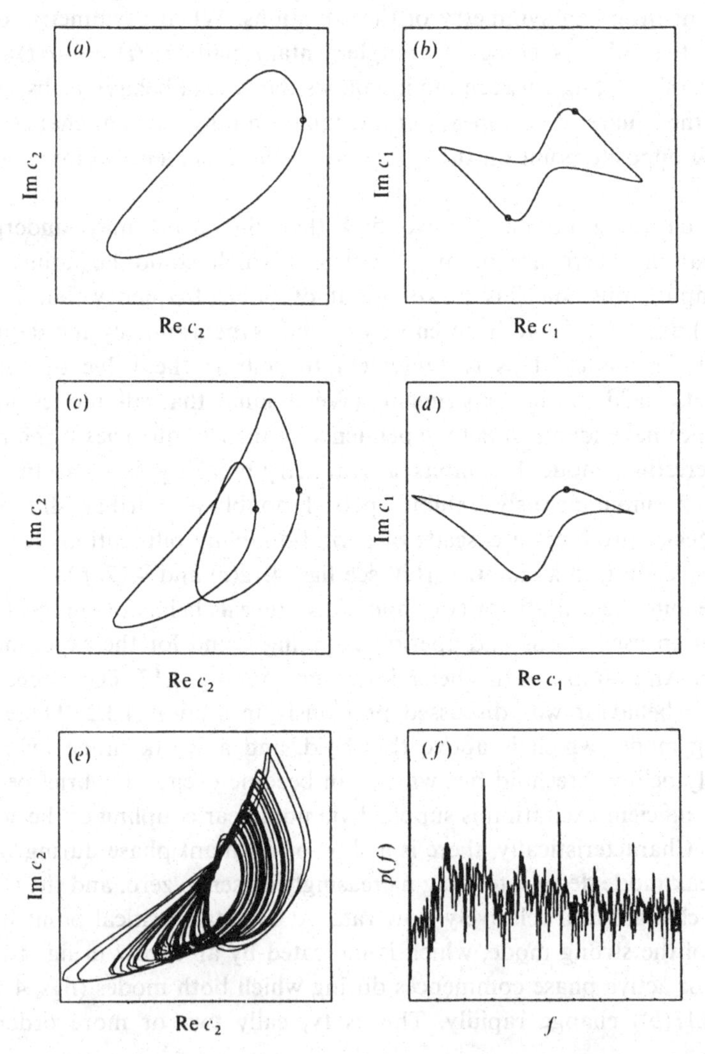

Table 4.4 *Parameter values used in figs. 4.12–14.*

P_{in}	2.7 $\begin{pmatrix}1.35 \text{ in fig. } 4.13, \\ 2.429 \text{ in figs. } 4.14(g), (h)\end{pmatrix}$ $(\times 10^{-2}\ \text{W})$
$\gamma_{\mathbf{k}}$	$1.0\ (\times 10^{6}\ \text{s}^{-1})$
$G_{\mathbf{k}}$	$1.414\ (\times 10^{7}\ \text{W}^{-1/2}\ \text{s}^{-1})$
$T_{\mathbf{k},\mathbf{k}'}$	$-1.896\delta_{\mathbf{k},\mathbf{k}'}\ (\times 10^{19}\ \text{G}^{-2}\ \text{s}^{-2})$
$S_{\mathbf{k},\mathbf{k}'}$	4.078 (3.971($\mathbf{k} = \mathbf{k}'$), 4.265($\mathbf{k} \neq \mathbf{k}'$) in figs. 4.14($a$)–($d$), ($e$), ($h$)) $(\times 10^{19}\ \text{G}^{-2}\ \text{s}^{-2})$
$U_{\mathbf{k},\mathbf{k}'}$	0

appearance as in fig. 4.13(d) (compare this with experimental results in fig. 4.9(e)), while orbits with a long dormant phase may tend to have a 'spiking' appearance, see fig. 4.13(c). This dormant period may become arbitrarily long for certain parameter values. This is because the orbit is approaching a saddle loop or homoclinic bifurcation when it contacts the saddle point at the origin. Beyond this point a transition must occur to another attractor – typically, a nonzero fixed point for the strong mode and zero for the weak mode. It should also be noted that relaxation and spiking behaviors do not have to be irregular – they may, for appropriate parameter values, be perfectly periodic. In some cases a cascade of period-doubling bifurcations to chaos has been observed to occur over extremely small changes in parameters ($<0.1\%$ change). This has the appearance, on first examination, of an emergence of irregularity of the orbit starting at a critical parameter value.

(3) Three-modes dynamics: this case yields other new phenomena. One of these is the emergence of quasiperiodic behavior with two incommensurate frequencies. An example of this is shown in fig. 4.14. The three modes do not have exactly the same frequency, but rather are spaced equally by a small separation in frequency to simulate the effect of a series of modes as observed experimentally. Naturally, all these frequencies must be very near to half of the pumping frequency, and it is the detuning which plays an important role in the dynamics. The quasiperiodic orbit lies on a two-torus in phase space. By strobing every cycle we can make a Poincaré section of the orbit. For a quasiperiodic orbit below the

transition to chaos, these points all lie on a closed curve – the intersection of the two-torus with the surface of section. The section may be defined in various ways; in fig. 4.14(*a*) the section points are the maximum value of Im c_1 for each cycle. The orbit shown can be found to emerge from a simple periodic orbit by a Hopf bifurcation. In the Poincaré section, the periodic orbit appears as a single point. This point spawns a circle at the

Fig. 4.13 (*a*), (*b*) Computed trajectories with Poincaré section marked by circles. Origin is at center of figure as indicated. (*c*), (*d*) Time series. (*e*) Power spectrum for (*d*). In (*a*)–(*c*) $\Delta f_1 = -200$ kHz, $\Delta f_2 = 300$ kHz; (*d*), (*e*) $\Delta f_1 = -175$ kHz, $\Delta f_2 = 325$ kHz. Other parameters values are given in table 4.4.

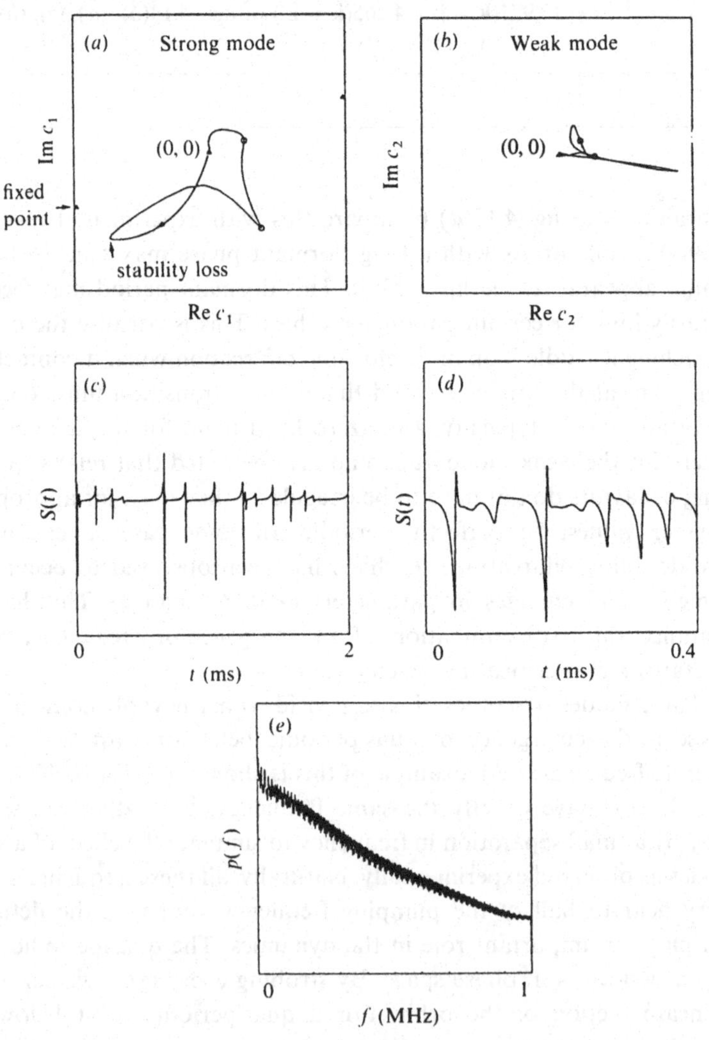

Fig. 4.14 (*a*), (*b*) and (*g*) Computed trajectories with Poincaré section marked by circles. (*c*), (*d*) and (*f*) Power spectra for (*a*), (*b*) and (*e*), respectively. (*e*) Poincaré section. (*h*) Time series for (*g*). In (*a*)–(*d*) $\Delta f_1 = -336$ kHz, $\Delta f_2 = 164$ kHz, $\Delta f_3 = 664$ kHz; (*e*), (*f*) $\Delta f_1 = -334.5$ kHz, $\Delta f_2 = 165.5$ kHz, $\Delta f_3 = 665.5$ kHz; (*g*), (*h*) $\Delta f_1 = -300$ kHz, $\Delta f_2 = 200$ kHz, $\Delta f_3 = 700$ kHz. Other parameter values are given in table 4.4.

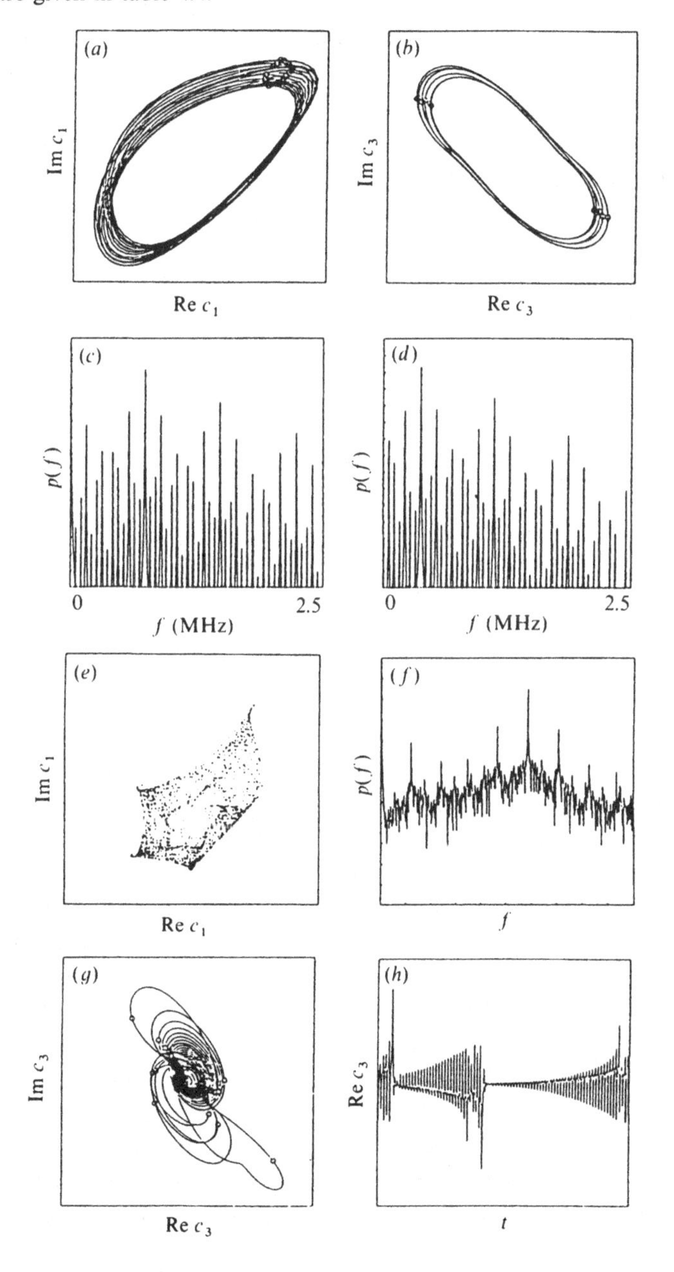

bifurcation point which initially grows in size in proportion to the square root of the change in parameter from its value at bifurcation. While mode 1 and mode 2 (not shown) exhibit asymmetric orbits, mode 3 has a symmetric quasiperiodic orbit as shown in fig. 4.14(*b*). Section points on this orbit are made simultaneously with those of mode 1. They occur twice each cycle because the basic period here is twice that of mode 1. The symmetry causes a restriction on frequencies that appear in the spectrum. The spectrum for mode 1 is shown in fig. 4.14(*c*). Here the allowed frequencies are all two-component harmonics of form $f_{mn} = mf_1 + nf_2$. The choice of f_1 and f_2 is not unique, but it is generally preferable to choose the two highest peaks. Figure 4.14(*d*) shows the spectrum for mode 3. Here the allowed peaks are those for which $m + n$ is odd. This can be shown to result from the symmetry of the orbit.

By changing the parameter values away from the Hopf bifurcation point, the two-torus on which the orbit lies grows larger and becomes less smooth. At a certain critical point the orbit may become chaotic and the torus becomes fractal. This is the quasiperiodic route to chaos. In figs. 4.14(*e*) and 4.14(*f*) are shown a Poincaré section of such a chaotic orbit and its power spectrum. This was reached by shifting the frequencies of the three modes synchronously, to simulate the effect of shifting the d.c. magnetic field in the experiment. The orbit is near to a period-5 phase locking, as can be seen in the five-pointed character of the section and in the spectrum. There remain some surprisingly sharp peaks in the spectrum considering the complexity shown in the Poincaré section.

More complicated versions of relaxation oscillations are sometimes seen for three modes (as well as the variety previously described). Figures 4.14(*g*) and 4.14(*h*) show a case in which only the third mode is of the 'weak' variety, while the other two modes are normally oscillating at a high level. The excitation which the weak mode receives from the pump and from the 'strong' modes is just sufficient to allow it to grow at a slow rate from an initially very small amplitude. This growth may extend over several orders of magnitude in amplitude, lasting for a hundred cycles or more of strong mode oscillation. Finally, when the size of the orbit becomes comparable to the size of the strong mode orbits, a rapid interaction phase occurs, involving all three modes, which leads to the return of the weak mode to a very low amplitude, and then the process repeats (approximately). This appears to be an orbit of 'Silnikov' or spiral saddle type (Guckenheimer and Holmes, 1983), for which existence of 'horseshoes' and other complex behavior have been explicitly shown. Cascon *et al.* (1991) pointed out atypical features of 'Silnikov' orbits in spin-wave systems, noting that the homoclinic orbit should include another saddle point besides the saddle focus at the origin.

This representation of the system as a series of equally spaced modes may be taken to the limit of an infinite series. The reason that this procedure is valid is that only those modes with relatively small detuning can become excited and interact with other excited modes. Modes which decay to zero have no effect on excited modes. Thus we need to include in computer simulation only those modes in the series with sufficiently small detuning (positive or negative). The necessary number may be found by extending the series by one mode at a time until the new modes added are observed to be inactive. Since the series is infinite, its behavior in parameter space is periodic with respect to synchronously shifting the frequencies by the mode spacing. In fig. 4.15, one period of the computed parameter space diagram is shown, giving mode frequency shift f_s versus applied power P_{in} (shift corresponds to change of d.c. field in experiment). The dynamics involve active participation of zero, one, two, or three modes. At higher powers than shown in the figure, additional modes may become involved. We label the modes as follows: mode 1 has

Fig. 4.15 Computed parameter-space diagram for mode series.

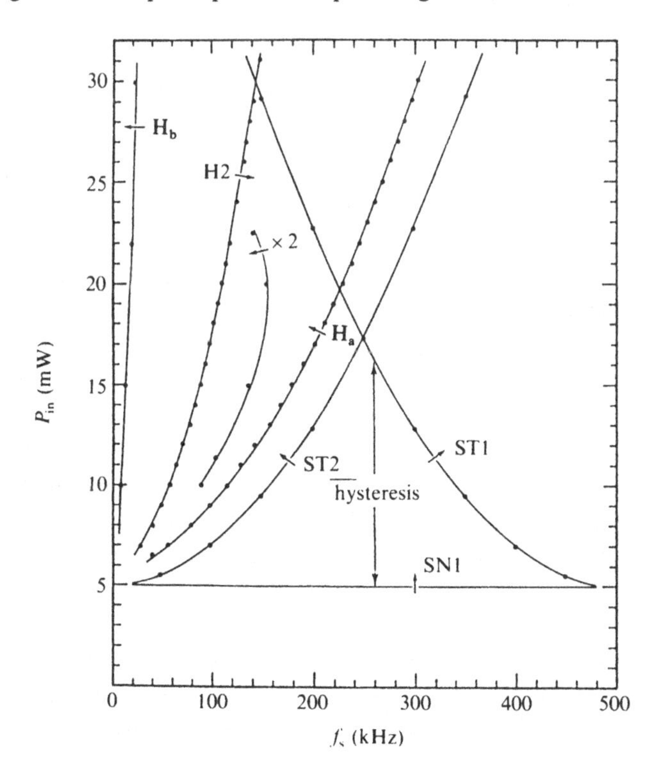

$\Delta f_1 = f_s - 500$ kHz, mode 2 has $\Delta f_2 = f_s$, mode 3 has $\Delta f_3 = f_s + 500$ kHz. ST1 and ST2 are the Suhl thresholds for excitation of modes 1 and 2, respectively, when all other modes are set to zero (this is actually a symmetry-breaking bifurcation for the stable fixed point at zero). SN1 is a saddle–node bifurcation of nonzero fixed points below Suhl threshold, see fig. 4.10(*e*). Hysteresis is observed when traversing the region between SN1 and ST1. Crossing H_a in the direction of the arrow, a Hopf bifurcation occurs in which a limit cycle involving modes 1 and 2 emerges from a fixed point. This is of the type shown in figs. 4.12(*a*) and 4.12(*b*). This oscillation undergoes a period-doubling bifurcation upon crossing the line labeled $\times 2$. Beyond this a cascade of period doublings occurs on route to chaos. On approach to the line H2, mode 3 becomes active and the two-mode solution we have been following is abruptly lost. H2 actually corresponds to a secondary Hopf bifurcation from a three-mode periodic orbit (on the left-hand side) to a three-mode quasiperiodic orbit (on the right-hand side), of the type shown in fig. 4.14. H_b corresponds to another primary Hopf bifurcation, this time involving modes 2 and 3. Relaxation oscillations occur above line ST1 in the section below its intersection with ST2. Onset appears to occur at Suhl threshold ST1. All these features are summarized in fig. 4.15.

4.4 Strange attractors and multifractals

Throughout sections 4.2 and 4.3, we have been concerned with spin-wave instabilities in first-order perpendicular pumping. In the case of parallel pumping (see section 4.1), period-doubling routes to chaos, etc. can also be observed.

Here we introduce a characterization (Mino and Yamazaki, 1986) of chaos of parallel-pumped magnons. Experiments were performed for a disk-shaped YIG (1.28 mm diameter and 0.04 mm thick) at a pumping frequency of 8.9 GHz at low temperature ($T = 4.2$ K). Both microwave and static fields are applied along the [111] direction perpendicular to the disk. Measurements were done in a field of 1935 G where the minimum threshold for instability is given by $P_{th} = 0.3$ mW, which will be taken as $P_{th} = 0$ dB in the following. The excited magnons have wavenumber $k \sim 0$ and propagate perpendicularly to the static field. Reflected microwave signals are detected by a tunnel diode and recorded in a computer at intervals of 1 μs.

Strange attractors are a set of phase-space trajectories in a chaotic state, where two nearby trajectories depart exponentially from one another with time. Exponential separation of trajectories forms a stretched sheet-like structure which will then be folded because of the bounded nature of phase space. In order to examine the three-dimensional character of the strange attractor, we have to generate three different sets of data. In experiments one obtains only single-time-series data for the number of specific parallel-pumped magnons. Those magnons are interacting with other magnon modes so as to cause auto-oscillations. Multidimensional trajectories might be constructed by the time-series data for several magnon modes. However, there is no way to measure separately a number of different magnon modes at the same time. The procedure of time delay is a conventional method of generating multidimensional data from the original single-time-series data. For example, three-dimensional time-series data $[\mathbf{V}(t), \mathbf{V}(t+\tau), \mathbf{V}(f+2\tau)]$ are generated by using a delay time τ (arbitrary), which forms trajectories embedded in three-dimensional phase space. For YIG $\tau = 3\ \mu s$ is a convenient value, which is about one-fifth of the fundamental period. Figure 4.16 shows a two-dimensional projection of a strange attractor on a $\mathbf{V}(t)$ versus $\mathbf{V}(t+\tau)$ plane. The axis vertical to the plane denotes $\mathbf{V}(t+2\tau)$. Figure 4.17 shows Poincaré sections given by intersection of positively directed trajectories with planes normal to the page passing through A–J of fig. 4.16. As is clearly seen in this figure, the trajectories form a two-dimensional sheet. By examining the evolution of trajectories in Poincaré sections at successive points along the strange attractor, the sequence of stretching (H → I → J → A → B) and folding (C → D → E → F → G → H) is evidently observed. An infinitely repeated stretching and folding process with evolution of trajectories gives a fractal structure proper to the strange attractor.

Fig. 4.16 Two-dimensional projection of strange attractor of YIG at $P = 3.60$ dB. (Courtesy of M. Mino and H. Yamazaki.)

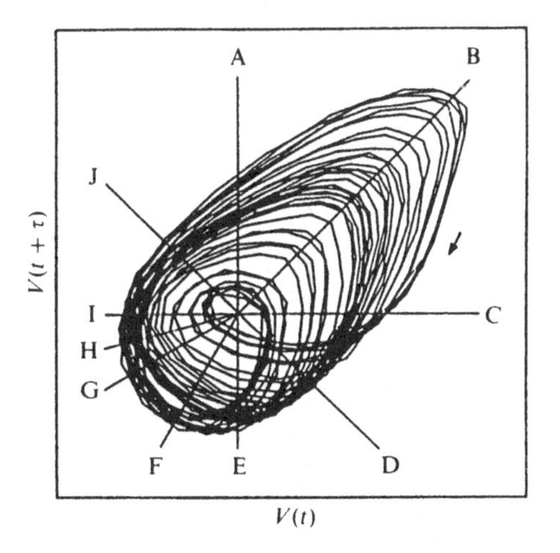

To compare experimental results with a theoretical model, detailed characterization of strange attractors is required. We proceed to describe an experimental study (Mino *et al.*, 1989) of strange attractors at and beyond the period-doubling accumulation point for a parallel-pumped spin-wave instability in YIG. In particular, change in multifractal structure of the strange attractor will be examined by increasing the driving power. For details of multifractals, refer to chapter 3.

Figure 4.18 shows power spectra and strange attractors which are constructed by the time-delay method ($\tau = 3\ \mu s$). An observed period-doubling cascade accumulates at $P = P_c = 2.20$ dB, and developed chaotic oscillations whose power spectra exhibit broad bands are obtained at $P = 2.97$ and 3.60 dB.

Singularity spectra $f(\alpha)$ are procured from complete trajectories embedded in five-dimensional space (each attractor has 20 000 points).

Fig. 4.17 Poincaré sections of strange attractor for planes normal to the paper passing through lines A–J of fig. 4.16. (Courtesy of M. Mino and H. Yamazaki.)

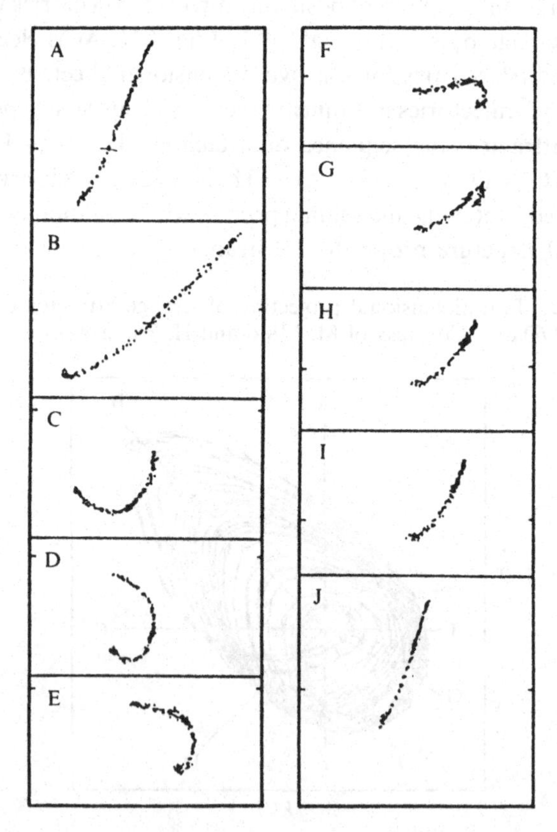

We first calculate the partition function $\Gamma(q, l) = \langle P_i(l)^{q-1} \rangle$ (see (3.57)–(3.62)), where the probability $P_i(l)$ is obtained by counting the number of points within the ith hyperspherical distribution with radius l and dividing it by the total number of points in the trajectory data set. The angular brackets represent an average over 2000 randomly chosen distributions. $\Gamma(q, l)$ has a wide scaling region satisfying $\Gamma(q, l) \sim l^{\tau(q)}$, which leads to $f(\alpha)$ by use of the Legendre transformation. Our experimental results have precision up to scale $l \sim 2^{-4}$, below which noise effects become operative. As the obtained $f(\alpha)$ spectra contain a trivial one dimension in the direction of trajectories, we hereafter subtract unity from both f and α.

Fig. 4.18 Experimental results for power spectra (*a*)–(*c*), and strange attractors (*d*)–(*f*): (*a*), (*d*) $P = P_c = 2.20$ dB; (*b*), (*e*), $P = 2.97$ dB; (*c*), (*f*) $P = 3.60$ dB.

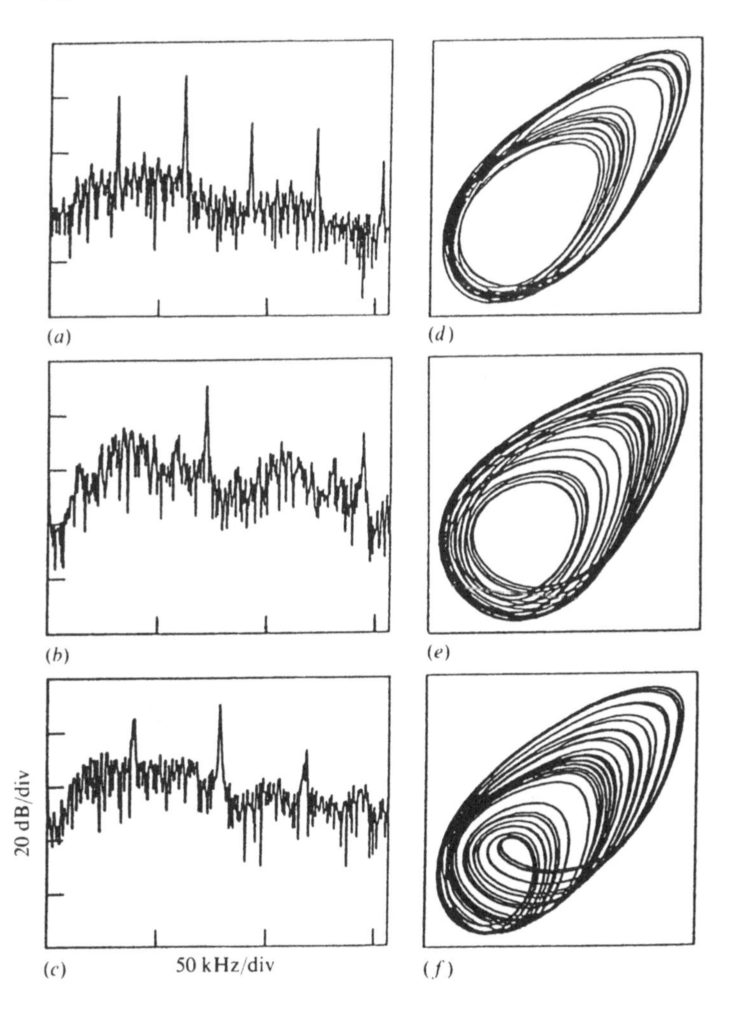

The $f(\alpha)$ spectrum at $P = P_c = 2.20$ dB is shown in fig. 4.19(*a*). The maximum point of the curve (i.e., D_0) is 0.55 ± 0.04. This curve is consistent with the universal one for the period-doubling route (see the solid curve in fig. 4.19(*a*)). At $P = 2.97$ dB, however, $\Gamma(q, l)$ has two distinctly different scaling regions separated by a crossover region R_c, see fig. 4.19(*d*). The scaling exponents $\tau(q)$ below and above R_c differ from each other. (Error bars associated with the individual scaling regions are too small to overcome the difference between these two exponents.) Consequently, there occur two kinds of coexisting humps or a 'bifurcation' of $f(\alpha)$, see fig. 4.19(*b*). This is due to band structures of strange attractors (i.e., island structures of their Poincaré sections) emerging from the period-doubling route. In fig. 4.19(*b*), the left-hand curve with $D_0 = 0.6$ is related to a larger-scale behavior (i.e., weak bunching, which wanders with different bands), retaining a feature of the universal curve at the critical point. The right-hand curve with $D_0 = 1.0$ is related to a small-scale curve (i.e., strong bunching in each band), describing a new multifractal structure. (The scales of both abscissas and ordinates in figs. 4.19(*b*) and

Fig. 4.19 Experimental $f(\alpha)$ spectra: (*a*) $P = 2.20$ dB; (*b*) $P = 2.97$ dB; (*c*) $P = 3.60$ dB. (*d*) Experimental $\Gamma(q, l)$ in logarithmic scales at $P = 2.97$ dB ($q = 0$).

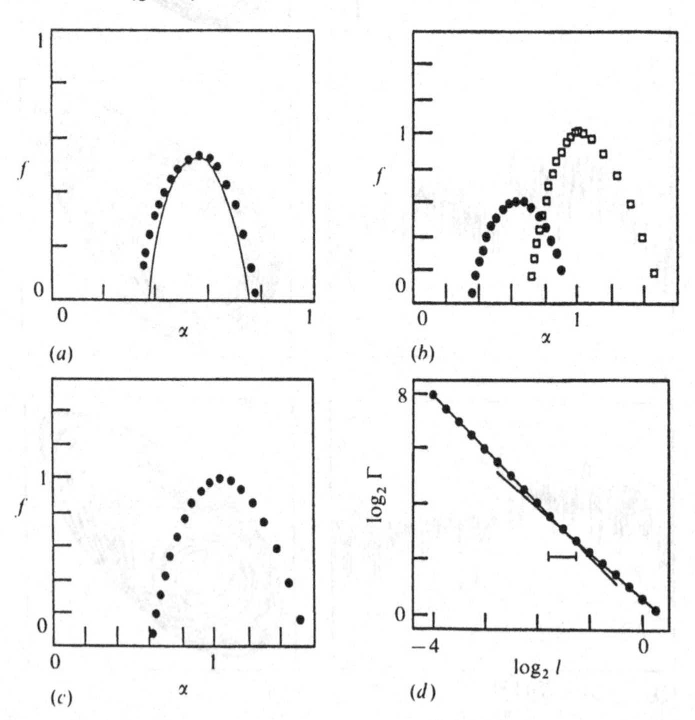

4.19(c) differ from those in fig. 4.19(a).) At $P = 3.60$ dB, a unique $\tau(q)$ is recovered.

The $f(\alpha)$ curve in fig. 4.19(c) has a fractal dimension $D_0 = 1.0$. By increasing power further, the oscillations become periodic again and exhibit period halving. Periods 4, 2 and 1 are observed at $P = 4.08$, 4.32 and 5.25 dB, respectively. A further increase of driving power causes growth of higher-dimensional attractors. The correlation dimensions D_2, which are obtained by use of a correlation integral, are 1.4, 2.0, 2.4 and 3.2 at $P = 8.56$, 13.01, 14.73 and 16.20 dB, respectively. Thus, bifurcation of $f(\alpha)$ is followed by growth of high-dimensional attractors. This kind of bifurcation is to be expected in the quasiperiodic route to chaos, because the circle map also shows a band structure in a sequence of Fibonacci ratios.

The theoretical model used in previous sections is again promising. The equation of motion for spin-wave modes $c_{\mathbf{k}}$ $(=c_{-\mathbf{k}})$ is

$$\dot{c}_{\mathbf{k}} = -\gamma_{\mathbf{k}} c_{\mathbf{k}} - \mathrm{i}\,\Delta\Omega_{\mathbf{k}} c_{\mathbf{k}}$$

$$- \mathrm{i}QFg_{\mathbf{k}}c^*_{\mathbf{k}} - \mathrm{i}\left(2\sum_{\mathbf{k}'} T_{\mathbf{k}\mathbf{k}'}|c_{\mathbf{k}'}|^2 c_{\mathbf{k}} + \sum_{\mathbf{k}'} (S_{\mathbf{k}\mathbf{k}'} + Eg_{\mathbf{k}}g^*_{\mathbf{k}'})c^2_{\mathbf{k}'}c^*_{\mathbf{k}}\right), \qquad (4.36)$$

where $\gamma_{\mathbf{k}}$, $\Delta\Omega_{\mathbf{k}}$, $g_{\mathbf{k}}$ and F $(\equiv h_{\mathrm{p}}\mathrm{V})$ are damping constants for $c_{\mathbf{k}}$, frequency shifts, coupling between the cavity mode and $c_{\mathbf{k}}$, and a driving field, respectively. $T_{\mathbf{k}\mathbf{k}'}$ and $S_{\mathbf{k}\mathbf{k}'}$ denote coupling between spin waves. Equation (4.36) is formally identical to the equation employed in the case of first-order perpendicular pumping. However, in the parallel-pumping case a cavity mode couples directly with spin-wave pairs, thereby producing new expressions for Q and E: $Q = -\mathrm{i}/\Gamma$ and $E = -\mathrm{i}/(2\Gamma)$, with Γ being the damping constant for the cavity mode. Further, in marked contrast to the perpendicular-pumping case, g_k terms vanish for modes propagating parallel to a static field.

We now confine our discussion to the case of a two-mode system where $g_{k1} \neq 0$ and $g_{k2} = 0$. Noting the realistic parameter values used in the previous section, we take $\hat{F} = F \times 10^{-7}\,\mathrm{s}^{-1}$ in the following. Consistent with experimental results, a period-doubling cascade accumulates at critical point $\hat{F} = \hat{F}_{\mathrm{c}} = 1.922\,86$. When $\hat{F}$ is increased beyond $\hat{F}_{\mathrm{c}}$, we see gradual growth of a higher-dimensional strange attractor. Some aspects of this dramatic change in the attractor may also be described on the basis of a unique low-dimensional map, e.g. a Hénon map. Our microscopic model has a great advantage, however, because it is directly related to a set of microscopic material constants of YIG. Singularity spectra $f(\alpha)$ are calculated here by using the Poincaré sections of the attractors. (We have obtained 8000 points for each section.)

The $f(\alpha)$ curve at $\hat{F} = \hat{F}_c$ in fig. 4.20(*a*) is almost identical to the universal one. The accuracies of D_0, α_{max} and α_{min} are indicated by error bars. At $\hat{F} = 1.9299$, $\Gamma(q, l)$ has a crossover region R_c (see fig. 4.20(*d*) and note that the noise level lies only in the region $l \leq 2^{-12}$), and 'bifurcation' of $f(\alpha)$ is observed, see fig. 4.20(*b*), which is quite similar to experimental results in fig. 4.19(*b*). When $\hat{F}$ increases further, R_c moves toward a larger scale region, and finally at $\hat{F} = 1.94$ in fig. 4.20(*c*), we find no indication of a crossover region. Corresponding D_0 values are 0.54 and 1.07 at $\hat{F} = \hat{F}_c$ and 1.94, respectively, while we have two D_0 values, 0.51 and 0.92, at $\hat{F} = 1.9229$. (A one-mode model here can, of course, yield no chaotic attractor and the two-mode model proves adequate to describe bifurcation of $f(\alpha)$.) Furthermore, we have obtained $D_0 = 1.8$ at $\hat{F} = 2.20$. Much larger values of D_0 will be available by increasing the number of active spin-wave modes. The theoretical model thus explains most of our experimental results very well. The experiments indicate a new route toward a high-dimensional attractor via the transition region where $f(\alpha)$ shows a remarkable bifurcation.

Fig. 4.20 Computed $f(\alpha)$ spectra of (4.36): (*a*) $\hat{F} = \hat{F}_c = 1.92286$; (*b*) $\hat{F} = 1.9229$; (*c*) $\hat{F} = 1.94$. (*d*) Computed $\Gamma(q, l)$ in logarithmic scales at $\hat{F} = 1.9229$ ($q = 0$).

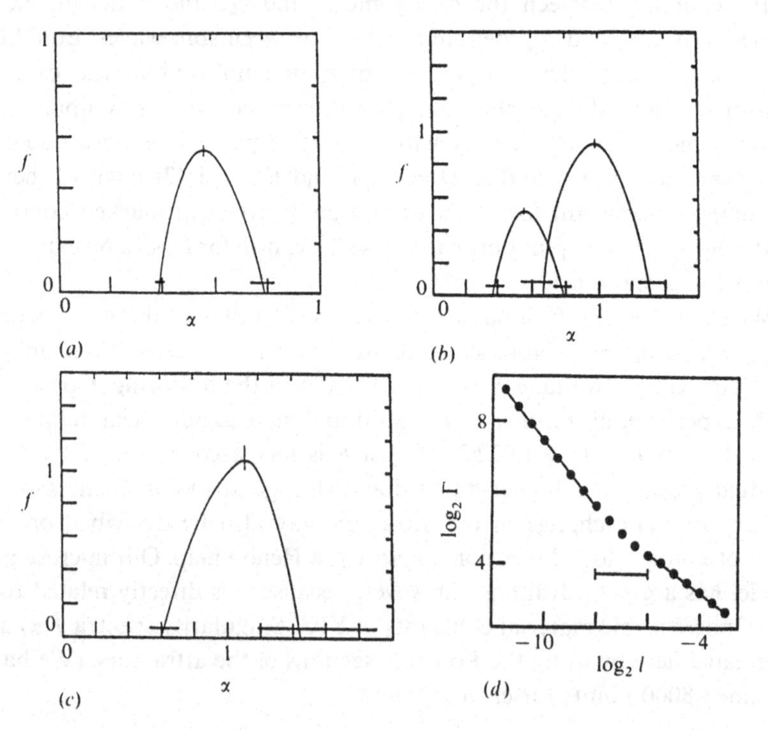

The macroscopic quanta of magnons have thus turned out to show universal routes to chaos, just as in fluid dynamics (e.g., Rayleigh–Benard convection and Taylor-vortex flows) and in Belousov–Zhabotinskiĭ chemical reactions. The most attractive feature of our 'solid-state turbulence' is that the microscopic interactions are well known. The microscopic theoretical model which includes the effect of a cavity mode explains very well a variety of experimental results (period-doubling cascade, quasi-periodicity, irregular relaxation oscillations and chaos). It mays also be possible to model features of quantum chaos. Then our theory should be modified so as to recover the quantum fluctuations lost in (4.22).

In closing this chapter other directions of current research may be mentioned. In Ga-doped YIG, Gibson and Jeffries (1984) observed period-doubling bifurcations, chaos and windows in the second-order perpendicular pumping case. Using the two-modes model employed by Nakamura *et al.* (1982), de Aguiar and Rezende (1986) attempted to explain the results of this experiment. In the parallel pumping case, materials other than YIG have been used. Waldner *et al.* (1985) observed spiking auto-oscillations and intermittency in two-dimensional antiferro-magnetic $(H_3NCH_2CH_2NH_3)CuCl_4$ and Yamazaki and Warden (1986) analyzed chaos in two-dimensional ferromagnetic $(CH_3NH_3)_2CuCl_4$ by measuring its fractal dimensions as a function of microwave power. Carroll *et al.* (1987) performed experiments on the first-order perpendicular pumping case with very low-frequency microwaves (2.0–3.3 GHz). Since in this case both ω_0 and $\omega_{\pm k}$ lie within the spin-wave band, instability occurs at very low pumping power. They analyzed quasiperiodicity-induced chaos by measuring the transient time required to get a stable strange attractor. A study on the number of its active modes is being performed by Wiese and Benner (1990). Research in many other directions, both theoretical and experimental, is also in progress.

5
Universal dynamical system behind quantum chaos: a single-parameter case

For quantum bounded systems whose classical versions are nonintegrable and chaotic, we derive coupled dynamical equations for both eigenvalues and eigenfunctions with a nonintegrability parameter τ taken as 'quasi-time'. These equations are shown to be a classical Calogero–Moser–Sutherland system with internal complex-vector space. Their complete integrability is shown and exact soliton solutions are given. It is explained how solitons adequately describe successive avoided crossings observed in the energy-level structure when a single parameter is varied. The final part constructs the statistical mechanics of a generalized Calogero–Moser system and thereby indicates a novel way to go beyond the traditional framework of random-matrix theory. This chapter constitutes the heart of my description.

5.1 A remarkable bridging between quantum chaos and nonlinear dynamics

We have so far observed salient aspects of quantum chaos (e.g., avoided crossings and multifractals) by taking typical examples experimentally accessible in condensed-matter physics. Multiple avoided crossings or repulsions of energy levels are indeed an essential mechanism for complicated spectra with a definite symmetry in quantum bound systems which are classically nonintegrable and chaotic. To understand statistical aspects of these spectra, much effort (McDonald and Kaufman, 1979; Bohigas *et al.*, 1984) has been devoted to applying results from random-matrix theory (Mehta, 1967). In particular, much effort has been concentrated on numerical studies fitting accumulated data to level-fluctuation properties predicted by Gaussian orthogonal or unitary ensembles. There is, however, no persuasive and admissible reason why random-matrix theory, which was developed to interpret excitation spectra of heavy nuclei, should be applicable to analysis of quantum chaos. The most fundamental problem is to derive complicated level arrangements and wavefunction patterns by investigation of the original quantum Hamiltonian, without having recourse to any ambiguous procedure whose validity is not certain.

In previous chapters, we have solved eigenvalue problems for Hermitian operator $H(\tau)$ and unitary operator $U(\tau)$ where τ represents hereafter a nonintegrability parameter (e.g., strength of magnetic fields). In analyses of quantum chaos, in general, we encounter eigenvalue problems of forms $H(\tau)|n\rangle = x_n|n\rangle$ and $U(\tau)|n\rangle = \exp(-ix_n)|n\rangle$ corresponding to autonomous and driven nonautonomous systems, respectively. Here we can naturally ask a question: if τ is regarded as 'quasitime' and both eigenfunctions $\{|n\rangle\}$ and eigenvalues $\{x_n\}$ as dynamical variables, what will be the resultant universal dynamical system? While Pechukas (1983) made the first attempt along these lines, his main concern was derivation of level-spacing distributions such as Wigner–Dyson or Poisson distributions. Therefore, he chose, besides variables $\{x_n\}$, matrix elements for a nonintegrable part of $H(\tau)$ as dynamical variables. The problem is that this methodology inevitably suppresses essential information possessed by eigenfunctions: in the case of finite N-dimensional Hilbert space, for instance, a set of eigenfunctions of $H(\tau)$ has N^2 complex variables, while nondiagonal matrix elements as employed by Pechukas have only $N(N-1)/2$ complex variables. Since aspects of quantum chaos are involved not only in energy spectra but also in wavefunctions, it is highly desirable to take both $\{|n\rangle\}$ and $\{x_n\}$ on the same footing.

We now derive equations of motion for eigenvalues and eigenfunctions (rather than matrix elements). These constitute a Calogero–Moser–Sutherland system with internal complex vector space, which has been proposed in other contexts of nonlinear dynamics. We thereby demonstrate a remarkable bridging between quantum chaos and nonlinear dynamics.

5.2 Generalized Calogero–Moser system

We first concentrate on autonomous nonintegrable systems. Let $H = H_0 + \tau V$ be the Hamiltonian for a quantum bound system which is classically nonintegrable. H_0 and V are a classically integrable part and a nonintegrable perturbation, respectively. Both H and H_0 are Hermitian, i.e., $H^\dagger = H$ and $H_0^\dagger = H_0$. Hence $V^\dagger = V$. A single parameter τ denotes the strength of nonintegrability. A manifold of definite symmetry is considered. Then we can assume discrete eigenvalues $\{x_n(\tau)\}$ to be nondegenerate, suppressing a negligible possibility of accidental degeneracies (von Neumann and Wigner, 1929; Arnold, 1978). Eigenfunctions $\{|n(\tau)\rangle\}$ are complex orthonormal and form a complete set. If we take τ as quasitime, equations of motion for $x_n(\tau)$, $p_n(\tau)$ $(\equiv V_{nn} \equiv \langle n(\tau)|V|n(\tau)\rangle)$, $|n(\tau)\rangle$, and $\langle n(\tau)|$ can be obtained from the time-independent Schrödinger equation

$$H(\tau)|n(\tau)\rangle = x_n(\tau)|n(\tau)\rangle \tag{5.1}$$

as (Nakamura and Lakshmanan, 1986)

$$dx_n/d\tau = p_n, \tag{5.2a}$$

$$dp_n/d\tau = 2 \sum_{m(\neq n)} V_{nm}V_{mn}(x_n - x_m)^{-1}, \tag{5.2b}$$

$$d|n\rangle/d\tau = \sum_{m(\neq n)} |m\rangle V_{mn}(x_n - x_m)^{-1}, \tag{5.2c}$$

$$d\langle n|/d\tau = \sum_{m(\neq n)} \langle m| V_{nm}(x_n - x_m)^{-1}. \tag{5.2d}$$

Equations (5.2*c*) and (5.2*d*) have been derived as follows: the derivative of (5.1) with respect to τ yields

$$V|n\rangle + H\,d|n\rangle d\tau = p_n|n\rangle + x_n\,d|n\rangle/d\tau. \tag{5.3}$$

Its left-hand side equals

$$\sum_m |m\rangle V_{mn} + \sum_m |m\rangle x_m \langle m|(d/n\rangle/d\tau)$$

$$= p_n|n\rangle + \sum_{m(\neq n)} |m\rangle V_{mn} + \sum_{m(\neq n)} |m\rangle x_m V_{mn}(x_n - x_m)^{-1}, \tag{5.4}$$

where

$$\langle m|(d|n\rangle/d\tau) = \begin{cases} V_{mn}(x_n - x_m)^{-1} & \text{for } m \neq n, \quad (5.5a) \\ 0 & \text{for } m = n \quad (5.5b) \end{cases}$$

are used. Noting that $x_n(\tau) \neq$ const $(\equiv 0)$, we obtain (5.2*c*) together with its complex conjugate (5.2*d*). (The equality (5.5*b*) is not correct in general. Instead, $\langle n|(d|n\rangle/d\tau) = i\phi_n(\tau)$ and its complex conjugate should be employed, where $\phi_n(\tau)$ is a τ-dependent real parameter. This correction seems to yield additional terms on the right-hand sides of (5.2*c*) and (5.2*d*). This problem can be resolved as follows. Let us define $\Phi_n(\tau)$ as $d\Phi_n/d\tau = \phi_n$. If we regard $\{|n\rangle \exp(-i\Phi_n)\}$ and their Hermitian conjugates as $\{|n\rangle\}$ and $\{\langle n|\}$, respectively, together with the corresponding redefinition of $\{V_{nm}\}$, then the original forms of (5.2*c*) and (5.2*d*) remain unaltered.) The remaining part of (5.2), i.e., (5.2*a*) and (5.2*b*), can be derived more simply by taking the first- and second-order derivatives with respect to τ of $\langle n(\tau)|H(\tau)|n(\tau)\rangle = x_n(\tau)$, respectively.

For the study below, it is convenient to rewrite (5.2) in a perfectly canonical formalism. Let us define Λ_{nm} as

$$\Lambda_{nm} = \mathrm{i}^{-1} V_{nm}(x_n - x_m), \tag{5.6}$$

which reduces to

$$\Lambda_{nm} = \langle n|\mathrm{i}^{-1}[H_0, V]|m\rangle \equiv \langle n|\Lambda|m\rangle. \tag{5.7}$$

The operator Λ is τ-independent and proves to be Hermitian. Let λ_0 be the lowest eigenvalue of Λ; then $\bar{\Lambda} \equiv \Lambda - \lambda_0 I$ becomes a τ-independent and non-negative Hermitian operator and therefore can be decomposed as

$$\bar{\Lambda} \equiv L^{\dagger}L. \tag{5.8}$$

Here L is an appropriate τ-independent operator which has a unique inverse L^{-1}. Λ_{nm} thus becomes

$$\Lambda_{nm} = \langle n|L^{\dagger}L|m\rangle + \lambda_0\delta_{nm}. \tag{5.9}$$

$\Lambda_{nn} = 0$ is self-evident from (5.6), which means that $\langle n|L^{\dagger}L|n\rangle = -\lambda_0$. By use of Λ_{nm} in (5.9), (5.2*b*) becomes

$$\mathrm{d}p_n/\mathrm{d}\tau = 2 \sum_{(m \neq n)} \langle n|L^{\dagger}L|m\rangle(x_n - x_m)^{-3}. \tag{5.2b'}$$

From (5.2*c*) and (5.2*d*), one has for $L|n\rangle$ and $\langle n|L^{\dagger}$:

$$\mathrm{d}(L/n\rangle)/\mathrm{d}\tau = -\mathrm{i} \sum_{m(\neq n)} L|m\rangle\langle m|L^{\dagger}L|n\rangle(x_n - x_m)^{-2}, \tag{5.2c'}$$

$$\mathrm{d}(\langle n|L^{\dagger})/\mathrm{d}\tau = \mathrm{i} \sum_{m(\neq n)} \langle n|L^{\dagger}L|m\rangle\langle m|L^{\dagger}(x_n - x_m)^{-2}. \tag{5.2d'}$$

The singular feature of repulsive forces in (5.2*b*′)–(5.2*d*′) originates in avoided crossings of quantum chaos. Equations (5.2*a*) and (5.2*b*′)–(5.2*d*′) should also apply to (classically) integrable systems with a suitable choice of parameter τ. Once a desymmetrized manifold is employed, however, such integrable systems can exhibit no avoided crossings and each level and eigenstate is affected predominantly by the tails of repulsive forces. If desymmetrization is not perfect, energy levels with different quantum numbers will show true crossings everywhere in the spectra. These true crossings will utterly invalidate the procedure leading to (5.2).

Equations (5.2c') and (5.2d'), though substantially the same as (5.2c) and (5.2d) (note the time-independence of L and $L^\dagger$ and the presence of their inverses), take advantageous forms for our study below. $L|n\rangle$ and $\langle n|L^\dagger$ are now read as complex dual vectors. Equations (5.2a) and (5.2b')–(5.2d') describe the dynamics for both eigenvalues and eigenfunctions (or more precisely, their modification) and suggest a perfectly canonical formalism. A brief comment should be made here: equations (5.2c') and (5.2d'), when combined, yield

$$\begin{aligned} d(\mathrm{i}\Lambda_{nm})/d\tau &= \mathrm{i}\{[d(\langle n|L^\dagger)/d\tau]L|m\rangle + \langle n|L^\dagger[d(L|m\rangle)/d\tau]\} \\ &= \sum_{l(\neq m,n)} \Lambda_{nl}\Lambda_{lm}[(x_m - x_l)^{-2} - (x_n - x_l)^{-2}]. \end{aligned} \tag{5.10}$$

The result in (5.10), together with (5.2a) and (5.2b'), was derived approximately by Pechukas (1983) and later without approximations by Yukawa (1985).

We now show that (5.2a) and (5.2b')–(5.2d') are also available by a quite different approach. An effective 'classical' N-particle Hamiltonian is introduced with internal complex-vector space in N dimensions for each particle:

$$\mathscr{H}_N = \sum_{n=1}^{N} p_n^2/2 + 2^{-1}\sum_{n=1}^{N}\sum_{\substack{m=1\\(m\neq n)}}^{N} \langle n|L^\dagger\cdot L|m\rangle\langle m|L^\dagger\cdot L|n\rangle(x_n - x_m)^{-2}, \tag{5.11}$$

with $\langle n|L^\dagger\cdot L|m\rangle$ for $n \neq m$ defined as a scalar product of complex vectors $\langle n|L^\dagger$ and $L|m\rangle$. $\langle n|L^\dagger\cdot L|n\rangle = 0$ is assumed here at the outset. Applying Poisson brackets given by

$$\begin{aligned} \{A, B\} &\equiv \sum_n \left(\frac{\partial A}{\partial p_n}\frac{\partial B}{\partial x_n} - \frac{\partial A}{\partial x_n}\frac{\partial B}{\partial p_n}\right) \\ &+ \sum_n \left(\frac{\partial A}{\partial \langle n|L^\dagger}\frac{\partial B}{\partial \mathrm{i}L|n\rangle} - \frac{\partial A}{\partial \mathrm{i}L|n\rangle}\frac{\partial B}{\partial \langle n|L^\dagger}\right), \end{aligned} \tag{5.12}$$

one has canonical equations from (5.11) as

$$\begin{aligned} dx_n/d\tau &= \{\mathscr{H}_N, x_n\} = \partial\mathscr{H}_N/\partial p_n, \\ dp_n/d\tau &= \{\mathscr{H}_N, p_n\} = -\partial\mathscr{H}_N/\partial x_n, \\ d(\mathrm{i}L|n\rangle)/d\tau &= \{\mathscr{H}_N, \mathrm{i}L|n\rangle\} = \partial\mathscr{H}_N/\partial\langle n|L^\dagger, \\ d(\langle n|L^\dagger)/d\tau &= \{\mathscr{H}_N, \langle n|L^\dagger\} = -\partial\mathscr{H}_N/\partial \mathrm{i}L|n\rangle. \end{aligned} \tag{5.13}$$

Despite an unusual appearance of the definition in (5.12), $\langle n|L^\dagger$ and $\mathrm{i}L|n\rangle$

are mutually canonically conjugate – satisfying $\{\langle n|L^\dagger, \mathrm{i}L|m\rangle\} = \delta_{n,m}$. (All theorems concerning Poisson brackets, e.g., Jacobi identity, prove to be unaltered, merely by our noting that $\{p_n\}$ and $\{\langle n|L^\dagger\}$ are on equal footing and likewise for $\{x_n\}$ and $\{\mathrm{i}L|n\rangle\}$.) Equations (5.11) and (5.13) can be identified with (5.2*a*) and (5.2*b*′)–(5.2*d*′), if the number N in the latter is finite. Clearly, this equivalence holds even in the limit $N = \infty$. Equations (5.11) and (5.13) are found to describe a Calogero–Moser N-particle system in $1 + 1$ dimensions with internal complex-vector space in N dimensions (see fig. 5.1), which happened to be proposed by Gibbons and Hermsen (1984) in a quite different context of nonlinear dynamics, though in the present case $\langle n|L^\dagger \cdot L|m\rangle$, $L|n\rangle$ and $\langle n|L^\dagger$ are defined in a much more intricate way. The number of degrees of freedom of the system, (5.13), is $N(N + 1)$, consisting of N for positions and momenta of N particles (levels) and N^2 for internal freedom (complex eigenstates) with which each particle (level) is endowed.

The universal (classical) dynamical system lying behind quantum chaos has been found to be described by (5.13) with (5.11) and (5.12). The eigenvalue problem (5.1) is now reduced to the initial value problem of (5.13). Complete integrability of (5.13) can be elucidated, merely by rewriting the novel findings of Gibbons and Hermsen in our notation. Let us define $N \times N$ matrices P, X, Γ and W as

$$\begin{aligned} P_{nm} &= \delta_{nm}p_n + (1 - \delta_{nm})\mathrm{i}\langle n|L^\dagger \cdot L|m\rangle(x_n - x_m)^{-1}, \\ X_{nm} &= \delta_{nm}x_n, \\ \Gamma_{nm} &= (1 - \delta_{nm})\mathrm{i}\langle n|L^\dagger \cdot L|m\rangle(x_n - x_m)^{-2}, \\ W &= (L|1\rangle, L|2\rangle, \ldots, L|n\rangle, \ldots, L|N\rangle), \end{aligned} \tag{5.14}$$

where $\{L|n\rangle\}$ are N-component column vectors. Further, we define $W^\dagger$ as Hermitian conjugate to W. A Lax representation of (5.13) is then written as

$$\begin{aligned} \mathrm{d}P/\mathrm{d}\tau &= [\Gamma, P], \\ \mathrm{d}X/\mathrm{d}\tau &= [\Gamma, X] + P, \\ \mathrm{d}W^\dagger/\mathrm{d}\tau &= \Gamma W^\dagger, \\ \mathrm{d}W/\mathrm{d}\tau &= -W\Gamma. \end{aligned} \tag{5.15}$$

Fig. 5.1 Generalized Calogero–Moser system.

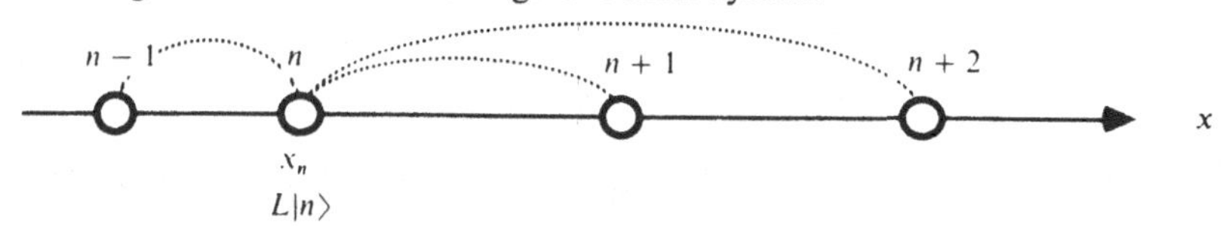

These matrix equations can be solved in a completely algebraic way:

$$\begin{aligned} X(\tau) &= U(\tau)[X(0) + P(0)\tau]U^{-1}(\tau), \\ P(\tau) &= U(\tau)P(0)U^{-1}(\tau), \\ W(\tau) &= W(0)U^{-1}(\tau), \\ W^{\dagger}(\tau) &= U(\tau)W^{\dagger}(0). \end{aligned} \tag{5.16}$$

Here $U(\tau)$ is a unitary matrix defined in terms of the time-ordering operator T, i.e.,

$$U(\tau) \equiv T\exp\left[\int_0^{\tau} \Gamma(\tau')\,d\tau'\right]. \tag{5.17}$$

(Note that $\Gamma^{\dagger}(\tau) = -\Gamma(\tau)$.) Lax forms and their solutions now serve to provide a wealth of knowledge of eigenvalues and wavefunctions. If at $\tau = +0$ (i.e., in the integrable or weak-coupling limit) values $\{x_n\}$, $\{p_n\}$, $\{|n\rangle\}$ and $\{\langle n|\}$ are given, the corresponding values at $\tau > 0$ (i.e., in the nonintegrable or strong-coupling regime) can be determined by use of (5.16).

There are three types of constant of motion: (i) $I_n = n^{-1}\,\mathrm{Tr}\,(P^n)$, (ii) $J_n(M^{(k)}) = \mathrm{Tr}\,(P^n W^{\dagger} M^{(k)} W)$ and (iii) $K_n = n^{-1}\,\mathrm{Tr}\,(W^{\dagger}W)^n$, where $n = 1, 2, \ldots, N$ and $M^{(k)}$ are diagonal traceless constant matrices with $k = 1, 2, \ldots, N-1$ which are linearly independent. The total number of constants of motion $N(N+1)$ is equal to the number of degrees of freedom of the system (5.13). These arguments suggest a new possibility of solving the puzzling problem of complicated spectra and wavefunctions in quantum chaos which has been studied mainly from a random-matrix theory viewpoint.

5.3 Generalized Sutherland system

Since the arguments in section 5.2 were limited to autonomous systems, the next and natural question is to look for an integrability behind nonintegrable, driven nonautonomous systems.

Let us consider a quantum Hamiltonian

$$H(t) = H_0 + \tau\hat{V} \sum_{j=-\infty}^{\infty} \delta(t - 2\pi j), \tag{5.18}$$

which describes any quantum bound system subjected to periodically pulsed field. H_0 and $\hat{V}$ correspond to classically integrable part and nonintegrable perturbation, respectively, and both of them are time (t)-independent Hermitian operators ($H_0^\dagger = H_0$, $\hat{V}^\dagger = \hat{V}$). Here τ denotes the strength of nonintegrability. For the time-dependent Schrödinger equation $i\hbar\, \mathrm{d}|\Psi\rangle/\mathrm{d}t = H(t)|\Psi\rangle$, its solution just after the jth pulse is given by $|\Psi_j\rangle = U^j|\Psi_0\rangle$. Here U is a one-period unitary operator defined in terms of time-ordering operator T as follows:

$$U \equiv U(\tau) = T\exp\left[\int_{+0}^{2\pi+0} (-\mathrm{i}/\hbar)H(t')\,\mathrm{d}t'\right]$$
$$= \exp(-\mathrm{i}\tau V)\,U_0, \tag{5.19}$$

where $V = \hat{V}/\hbar$ and $U_0 = \exp[(-\mathrm{i}/\hbar)2\pi H_0]$. So, the eigenvalue problem

$$U(\tau)|n(\tau)\rangle = \exp[-\mathrm{i}x_n(\tau)]|n(\tau)\rangle \tag{5.20}$$

and its quasienergies $\{x_n(\tau)\}$ (which are discrete because of the nature of bound spectra) and quasi-eigenfunctions $\{|n(\tau)\rangle\}$ determine the quantum dynamics. Let us consider a manifold with a definite symmetry. Then $\{x_n(\tau)\}$ can be assumed to be nondegenerate, by ignoring a negligible possibility of accidental degeneracies. $\{|n(\tau)\rangle\}$ are complex orthonormal and form a complete set. Taking τ as quasitime, equations of motion for $x_n(\tau)$, $p_n(\tau)$ $[\equiv V_{nn} \equiv \langle n(\tau)|V|n(\tau)\rangle]$, $|n(\tau)\rangle$, and $\langle n(\tau)|$ can be obtained from the time derivative of (5.20) as follows (Nakamura and Mikeska, 1987):

$$\mathrm{d}x_n/\mathrm{d}\tau = p_n, \tag{5.21a}$$

$$\mathrm{d}p_n/\mathrm{d}\tau = \mathrm{i}\sum_{m\neq n} V_{nm}V_{mn}(\{(1 - \exp[\mathrm{i}(x_m - x_n)]\}^{-1} - \text{c.c.}), \tag{5.21b}$$

$$\mathrm{d}|n\rangle/\mathrm{d}\tau = -\mathrm{i}\sum_{m\neq n} |\mathrm{m}\rangle V_{mn}\{1 - \exp[\mathrm{i}(x_n - x_m)]\}^{-1}, \tag{5.21c}$$

$$\mathrm{d}\langle n|/\mathrm{d}\tau = \mathrm{i}\sum_{m\neq n} \langle \mathrm{m}|V_{mn}\{1 - \exp[\mathrm{i}(x_m - x_n)]\}^{-1}. \tag{5.21d}$$

The derivation of the above equations is quite parallel to that in section 5.2: the τ-derivative of (5.20) yields

$$-\mathrm{i}VU|n\rangle + U\,\mathrm{d}|n\rangle/\mathrm{d}\tau = -\mathrm{i}p_n\exp(-\mathrm{i}x_n)\,|n\rangle + \exp(-\mathrm{i}x_n)\,\mathrm{d}|n\rangle/\mathrm{d}\tau. \tag{5.22}$$

Its left-hand side

$$= -\mathrm{i}\sum_m |m\rangle V_{mn} \exp(-\mathrm{i}x_n) + \sum_m |m\rangle \exp(-\mathrm{i}x_n)\langle m|(\mathrm{d}|n\rangle/\mathrm{d}\tau)$$

$$= -\mathrm{i}|n\rangle p_n \exp(-\mathrm{i}x_n) - \mathrm{i}\sum_{m\neq n} |m\rangle V_{mn} \exp(-\mathrm{i}x_n)$$

$$- \mathrm{i}\sum_{m\neq n} |m\rangle V_{mn}\{1 - \exp[\mathrm{i}(x_n - x_m)]\}^{-1} \exp(-\mathrm{i}x_m), \qquad (5.23)$$

where

$$\langle m|(\mathrm{d}|n\rangle/\mathrm{d}\tau) = -\mathrm{i}V_{mn}\{1 - \exp[\mathrm{i}(x_n - x_m)]\}^{-1} \qquad (5.24)$$

for $m \neq n$ together with $\langle n|(\mathrm{d}|n\rangle/\mathrm{d}\tau) = 0$ are used. Suppressing common term $-\mathrm{i}|n\rangle p_n \exp(-\mathrm{i}x_n)$, we obtain (5.21*c*) together with its complex conjugate (5.21*d*). (The problem concerning the equality $\langle n|(\mathrm{d}|n))\rangle/\mathrm{d}\tau) = 0$ has been resolved in the same way as in section 5.2.)

To elucidate the completely integrable nature of (5.21), it is convenient to rewrite these equations in a perfectly canonical form. Let us define Λ_{nm} as

$$\Lambda_{nm} = V_{nm}\{1 - \exp[\mathrm{i}(x_n - x_m)]\}, \qquad (5.25)$$

which reduces to

$$\Lambda_{nm} = \langle n|V|m\rangle - \langle n|U^{\dagger}VU|m\rangle$$
$$= \langle n|V - U_0^{\dagger}VU_0|m\rangle \equiv \langle n|\Lambda|m\rangle. \qquad (5.26)$$

The operator Λ is τ-independent and is found to be Hermitian. Therefore, the argument below (5.7) again holds: Λ can be written as

$$\Lambda = L^{\dagger}L + \lambda_0 I. \qquad (5.27)$$

Here L is a τ-independent operator and λ_0 is the lowest eigenvalue of Λ. By using (5.25) and (5.27), equation (5.21*b*) becomes

$$\mathrm{d}p_n/\mathrm{d}\tau = \mathrm{i}\sum_{m\neq n} \langle n|L^{\dagger}L|m\rangle\langle m|L^{\dagger}L|n\rangle|1$$
$$- \exp[\mathrm{i}(x_n - x_m)]|^{-2}(\{1 - \exp[\mathrm{i}(x_m - x_n)]\}^{-1} - \mathrm{c.c.})$$
$$= 4^{-1}\sum_{m\neq n} \langle n|L^{\dagger}L|m\rangle\langle|L^{\dagger}L|n\rangle \cos[(x_n - x_m)/2]$$
$$\times \sin^{-3}[(x_n - x_m)/2]. \qquad (5.21b')$$

In a similar way, (5.21*c*) and (5.21*d*) become

$$\mathrm{d}(L|n\rangle)/\mathrm{d}\tau = (-\mathrm{i}/4)\sum_{m\neq n} L|m\rangle\langle m|L^{\dagger}L|n\rangle \sin^{-2}[(x_n - x_m)/2], \qquad (5.21c')$$

$$\mathrm{d}(\langle n|L^{\dagger})/\mathrm{d}\tau = (\mathrm{i}/4)\sum_{m\neq n} \langle n|L^{\dagger}L|m\rangle\langle m|L^{\dagger} \sin^{-2}[(x_n - x_m)/2]. \qquad (5.21d')$$

Equations (5.21*a*) and (5.21*b*′)–(5.21*d*′) describe the dynamics for both quasienergies and quasi-eigenfunctions and take a perfectly canonical formalism. Let us introduce a 'classical' N-particle Hamiltonian with internal complex-vector space for each particle:

$$\mathscr{H}_N = \sum_{n=1}^{N} p_n^2/2 + 2^{-1} \sum_{n=1}^{N} \sum_{\substack{m=1 \\ (m \neq n)}} \langle n|L^\dagger \cdot L|m\rangle\langle m|L^\dagger \cdot L|n\rangle$$

$$\times 4^{-1} \sin^{-2}[(x_n - x_m)/2], \tag{5.28}$$

where $\langle n|L^\dagger \cdot L|m\rangle$ is now read as a scalar product of complex dual vectors. N can be either finite or infinite. Applying Poisson brackets similar to those in section 5.2, one has canonical equations from (5.28), which are found to be exactly the same as (5.21*a*) and (5.21*b*′)–(5.21*d*′). The 'classical' Hamiltonian in (5.28) describes a Sutherland many-particle system (Sutherland, 1971) generalized so as to include an internal complex-vector space for each particle. The number of degrees of freedom for the system in (5.28) is $N(N+1)$, consisting of N for position momenta of N particles (quasienergies) and N^2 for internal freedom (quasi-eigenfunctions).

We now show the astonishing fact that (5.21*a*) and (5.21*b*′)–(5.21*d*′), despite their complicated appearance, are again completely integrable. Let us define $N \times N$ matrices P, X, Γ and W as

$$\begin{aligned}
P_{nm} &= \delta_{nm} p_n + (1 - \delta_{nm})\mathrm{i}\langle n|L^\dagger \cdot L|m\rangle 2^{-1} \cot[(x_n - x_m)/2], \\
X_{nm} &= \delta_{nm} x_n, \\
\Gamma_{nm} &= (1 - \delta_{nm})\mathrm{i}\langle n|L^\dagger \cdot L|m\rangle 2^{-2} \sin^{-2}[(x_n - x_m)/2], \\
W &= (L|1,\rangle, L|2\rangle, \ldots, L|n\rangle, \ldots, L|N\rangle),
\end{aligned} \tag{5.29}$$

where $L|n\rangle$ are N-component column vectors. A Lax representation of (5.21) is then written as $\mathrm{d}X/\mathrm{d}\tau = \mathrm{diag}(P_{11}, P_{22}, \ldots)$, $\mathrm{d}P/\mathrm{d}\tau = [\Gamma, P]$, $\mathrm{d}W/\mathrm{d}\tau = -W\Gamma$, together with their Hermitian conjugates $\mathrm{d}P^\dagger/\mathrm{d}\tau = [\Gamma, P^\dagger]$ and $\mathrm{d}W^\dagger/\mathrm{d}\tau = \Gamma W^\dagger$. (Note that $\Gamma^\dagger = -\Gamma$.)

In marked contrast with autonomous systems in section 5.2, the spectrum $\{x_n(\tau)\}$ is periodic in energy with a period of 2π, as recognized in (5.20). In other words, the generalized Sutherland system that has been obtained is defined on a ring chain with length 2π. The constants of motion in involution are of three types. They take the same form as given below (5.17) with the new definition of P, X, Γ and W in (5.29). The validity of these constants of motion can be checked in a periodically pulsed quantum spin system (see chapter 3), where Hilbert space is in finite dimensions and all the bound quasienergies are calculable in a wide range of nonintegrability (i.e., magnetic field B). The spectrum in fig. 5.2 was obtained for the pulsed spin system (see chapter 3), by numerical diagonalization of (3.47) for various μB values (μ in the Bohr magneton multiplied by g factor ($=2.0023$)). Energies are depicted in a fundamental 'Brillouin' zone. We can see the absence of any accidental degeneracy in fig. 5.2. Then, by examining this figure together with numerical data for matrix elements, it is found that I_1(total momentum), $I_2 + 4^{-1}K_2$(total energy), etc., do not change, irrespective of change in the μB value. (An analogous study for a kicked quantum rotor whose Hilbert space is

Fig. 5.2 Field-dependent quasienergy diagram for a pulsed quantum spin system with spin magnitude $S = 16$ in chapter 3. Manifold with an even parity is depicted in fundamental zone ($0 \leq E/\hbar \leq 1$). $2\pi E/\hbar$ and μB (scaled by the easy-plane anisotropy energy) should be read as x and τ, respectively, in the present text.

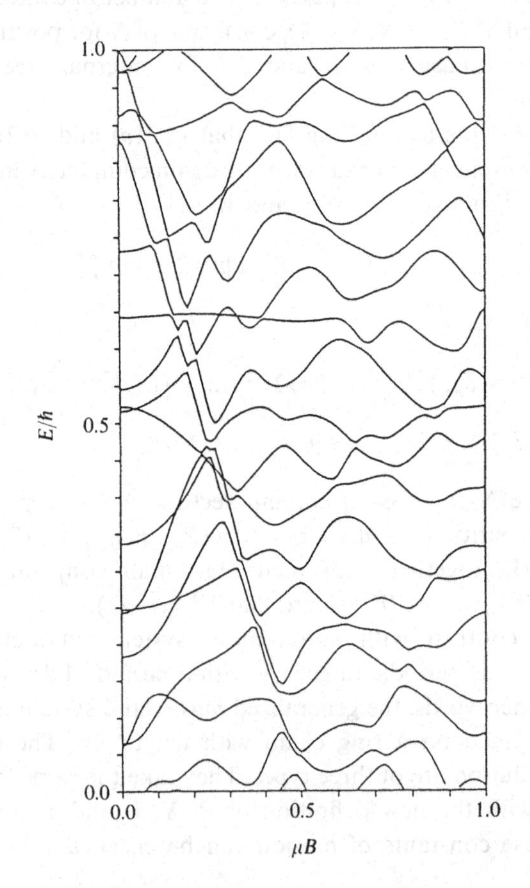

infinite dimensional will be done. However, inevitable matrix-truncation procedures will make us confirm the presence of constants of motion less rigorously.)

Constants of motion as described above will play a vital role in quantitative descriptions of quantum recurrence and of other complicated dynamics (see chapter 3). Since both Calogero–Moser and Sutherland systems without complex-vector space are typical examples of a completely integrable classical particle system with interparticle interactions of doubly-periodic Weierstrass function type $\mathscr{P}(z|\omega, \omega')$ (note, e.g., $\mathscr{P}(z|\infty, \infty) = z^{-2}$ and $\mathscr{P}(z|0, i\pi/2) = 4^{-1}\sin^{-2}(z/2)$) (Olshanetsky and Perelomov, 1981), the latter system with internal complex-vector space will be indicated as a more universal dynamical system which underlies quantum chaos.

In closing this section, two remarks should be added. First, while perturbations used in (5.1) and (5.18) were assumed nonintegrable, the scenarios leading to (5.11)–(5.13), (5.21*a*) and (5.21*b*′)–(5.21*d*′) hold well, irrespective of the integrability or nonintegrability of the perturbations, if the desymmetrization is made in advance. Readers are referred to the comment below (5.2*d*′). Second, while we have employed Hermitian $H(\tau)$ in (5.1), there exist also its orthogonal, unitary and symplectic variants. For real quaternions $H(\tau) = H_0 + \tau \sum_{a=0}^{3} \mathrm{e}_a V^{(a)}$ in general, elements Λ_{nm} in (5.7) and (5.26) are also generalized to real quaternions given by

$$\begin{aligned}\Lambda_{nm} &\equiv \sum_{a=0}^{3} \mathrm{e}_a \langle n|\mathrm{i}^{-1}[H_0, V^{(a)}]|m\rangle \\ &\equiv \mathrm{e}_0\Lambda_{nm}^{(0)} + \mathrm{e}_1\Lambda_{nm}^{(1)} + \mathrm{e}_2\Lambda_{nm}^{(2)} + \mathrm{e}_3\Lambda_{nm}^{(3)}. \end{aligned} \tag{5.30}$$

Here e_0 and e_a $(a = 1, 2, 3)$ are defined by

$$\mathrm{e}_0^2 = \mathrm{e}_0, \qquad \mathrm{e}_a^2 = -\mathrm{e}_0; \tag{5.31a}$$

$$\mathrm{e}_0\mathrm{e}_a = \mathrm{e}_a\mathrm{e}_0 = \mathrm{e}_a, \qquad \mathrm{e}_a\mathrm{e}_b = -\mathrm{e}_b\mathrm{e}_a = \mathrm{e}_c; \tag{5.31b}$$

$$\mathrm{e}_0^\dagger = \mathrm{e}_0, \qquad \mathrm{e}_a^\dagger = -\mathrm{e}_a. \tag{5.31c}$$

or, in terms of the Pauli matrices, given by $\mathrm{e}_1 = \mathrm{i}\sigma_z$, $\mathrm{e}_2 = -\mathrm{i}\sigma_y$, $\mathrm{e}_3 = -\mathrm{i}\sigma_x$ and $\mathrm{e}_0 = 1$ (the 2×2 unit matrix). Because Λ is Hermitian, $\Lambda_{nm}^{(0)}$ is real symmetric while $\Lambda_{nm}^{(a)}$ $(a = 1, 2, 3)$ are real antisymmetric with vanishing diagonal elements. For orthogonal and unitary cases, Λ_{nm} is real ($\Lambda_{nm}^{(a)} = 0$ for $a = 1, 2, 3$) and complex($\Lambda_{nm}^{(a)} = 0$ for $a = 2, 3$), respectively, whereas $\Lambda_{nm}^{(a)}$ $(a = 1, 2, 3)$ are nonvanishing for the symplectic case. These arguments also apply to the scalar products $\langle n|L^\dagger \cdot L|m\rangle$ in the corresponding classical dynamical systems in (5.11) and (5.28), where the generalization $L^+L \to \sum_{a=0}^{3} L^{(a)+} \cdot L^{(a)}\mathrm{e}_a$ should be made. Expression (5.30) will be employed in later sections.

5.4 Solitons and moving avoided crossings

We have now reached a stage at which we can give a novel insight into avoided crossings. A multitude of avoided crossings displayed in energy spectra is a clear signature of nonintegrability and chaos (see chapters 2 and 3). Successive avoided crossings along a fictive curve are also observed, suggesting persistence of some property of an eigenstate after several avoided crossings. We try to interpret the nature of avoided crossings from a nonlinear dynamics viewpoint (Nakamura, 1989; Gaspard *et al.*, 1989; Rice *et al.*, 1992).

We have already arrived at the conclusion that there exists a universal 'classical' dynamical system underlying quantum chaos. The systems in (5.11) and (5.28) have been shown to be completely integrable when the number of particles is finite. Accordingly, they share the rare feature of integrability with the finite ideal gas, harmonic chain and Toda chain. Moreover, (5.11) and (5.28) reduce to the Hamiltonian for a finite ideal gas when $\langle n|L^\dagger \cdot L|m\rangle = 0$ for any n and m $(n \neq m)$. When infinite extensions of their domains are assumed, these latter systems are known to support persistence modes of propagation. In particular, solitons can propagate in the infinite nonlinear Toda lattice. We further develop this analogy by giving here one-soliton and two-soliton solutions of the generalized Calogero–Moser (gCM) system in (5.11).

In order to construct such solutions, we note that each solution of system (5.11) corresponds to a parametric family of Hamiltonians of the form $H_0 + \tau V$ and vice versa. We therefore consider the following Hamiltonian operator (which has a long history (Fano, 1961; Bohr and Mottelson, 1953))

$$H(\tau) = \begin{bmatrix} p\tau & u_1 & u_2 & u_3 & \dots & u_N \\ u_1 & y_1 & 0 & 0 & \dots & 0 \\ u_2 & 0 & y_2 & 0 & \dots & 0 \\ u_3 & 0 & 0 & y_3 & \dots & 0 \\ \cdot & \cdot & \cdot & \cdot & \dots & \cdot \\ \cdot & \cdot & \cdot & \cdot & \dots & \cdot \\ \cdot & \cdot & \cdot & \cdot & \dots & \cdot \\ u_N & 0 & 0 & 0 & \dots & y_N \end{bmatrix} \tag{5.32}$$

where $y_n < y_{n+1}$. The eigenvalues of (5.32) are given by the roots of the equation

$$p\tau = x - \sum_{n=1}^{N} \frac{u_n^2}{x - y_n}. \tag{5.33}$$

This equation possesses $N + 1$ roots $\{x_n(\tau)\}$ which generate an exact solution of the system (5.11). The eigenvalue problem for (5.32) is thus identical to solving dynamical system (5.11) with initial values determined from the roots of (5.33) at $\tau = 0$. The energy spectrum is composed of N horizontal levels (i.e., independent of τ) which are crossed by one extra level with a slope given by the parameter p. In fact, each and every crossing is avoided. Repulsion between levels is controlled by coupling parameters $\{u_n\}$. For equal spacing between horizontal levels, $y_n = an$, the succession of avoided crossings corresponds to a soliton propagating with velocity p through a lattice of period a, if horizontal levels are extended for positive and negative values of coordinate x. The exact one-soliton solution is then given by the zeros $\{x_n(\tau)\}$ of

$$p\tau = x - \frac{\pi u^2}{a} \cot \frac{\pi x}{a}. \tag{5.34}$$

This formula gives us the soliton profile. At a large distance from the soliton, the decrease of its profile is algebraic, $x_n(\tau) \simeq an + u^2(an - p\tau)^{-1}$. For the one-soliton solution (5.34), the velocity of particles is

$$p_n(\tau) = p\left\{1 + \frac{\pi^2 u^2}{a^2}\left[\sin \frac{\pi x_n(\tau)}{a}\right]^{-2}\right\}^{-1}, \tag{5.35}$$

and the matrix elements of the interaction potential V are then (note (5.6) and (5.9))

$$\langle m(\tau)|V|n(\tau)\rangle = [p_m(\tau)p_n(\tau)]^{1/2}. \tag{5.36}$$

The spin and cospin can be obtained as explained above. The velocity (5.35) decreases at a large distance as

$$p_n(\tau) \simeq pu^2|an - p\tau|^{-2}. \tag{5.37}$$

Accordingly, the gCM soliton has long-range tails in contrast to the Toda soliton. The energy of a single soliton is given by

$$E_{\text{sol}} \simeq p^2u^2/2a^4. \tag{5.38}$$

Solitons are known to collide with each other without undergoing a change of structure. It is thus important to construct the bisoliton solution of (5.11) in order to verify this property. The family of Hamiltonian operators we need to consider for this purpose is

$$H(\tau) = \begin{bmatrix} q\tau + b & w & v & v & v & \dots \\ w & p\tau & u & u & u & \dots \\ v & u & a & 0 & 0 & \dots \\ v & u & 0 & 2a & 0 & \dots \\ v & u & 0 & 0 & 3a & \dots \\ \cdot & \cdot & \cdot & \cdot & \cdot & \dots \\ \cdot & \cdot & \cdot & \cdot & \cdot & \dots \\ \cdot & \cdot & \cdot & \cdot & \cdot & \dots \end{bmatrix}, \tag{5.39}$$

where u (v) is the coupling between soliton level with velocity p (q) and lattice levels, and w is the coupling between two-soliton levels. The solution is now given by the following quadratic equation in τ:

$$(q\tau + b - x)(p\tau - x) + f(x)[u^2(q\tau + b - x) + v^2(p\tau - x)] - 2uvwf(x) - w^2 = 0 \tag{5.40}$$

with $f(x) = (\pi/a)\cot(\pi x/a)$ in the limit of infinite extension. The bisoliton solution is then given by the real roots of (5.40). Figure 5.3 depicts an example of crossing between two solitons, which illustrates their stability with respect to collision.

Multisolitons can be constructed by generalization of the above procedure. When many solitons are created in this way, the spectrum appears irregular for any fixed value of parameter τ. We may then associate an irregular spectrum with an ideal gas of solitons: excitations of small and large number of solitons would correspond to near-integrable and chaotic regions, respectively. This idea is analogous to that exploited in (classical) fluid dynamics where some aspects of turbulence can be described in terms of a gas of vortices.

In the energy spectra of quantum chaos, we can actually recognize soliton-like structures. In fig. 5.2, downward propagation of local distortion in the low-field region may be interpreted as a soliton. The experiment by Iu *et al.* (1989) on diamagnetic Rydberg atoms (Li) in a magnetic field provides a more interesting energy spectrum (see fig. 5.4) including soliton-like structures in the background of orderly level structures. The regular structures in fig. 5.4 are amazing because they appear in the regime where the corresponding classical motion of electrons is fully chaotic. This fact reflects the presence of a small fraction of KAM tori within the ergodic sea in classical phase space.

5.5 Statistical mechanics of generalized Calogero–Moser system

In previous sections we have observed that the eigenvalue spectrum is assimilated to a gas of particles moving with quasitime parameter τ. Collisions between particles correspond to avoided level crossings. Since the Hamiltonian operator in (5.1) works in infinite-dimensional Hilbert space, we inevitably have an infinite number of energy levels. This fact encourages us to study the statistical properties of energy levels.

Let us consider, for example, the quasienergy spectrum for the pulsed-spin system in chapter 3. For spin-magnitude $S = 16$, we have the spectrum in fig. 5.2. When the spin magnitude is increased further, the number of energy levels will grow, rendering the spectrum more and more complicated. We acquire, however, the advantage of being able to formulate a statistical description. Figures 5.5 and 5.6 are the computed distributions of level spacings (5.67) and level curvatures (5.94), respectively. A multitude of avoided level-crossings can be seen to affect both distributions crucially: fig. 5.5(c) shows the characteristic increase of the

Fig. 5.3 Example of a bisoliton: energy spectrum of Hamiltonian family (5.39) with parameter values $p = 1$, $q = -2$, $u = 0.2$, $v = 0.3$, $w = 0.4$, $a = 1$, $b = 21$ and $n = 1, 2, \ldots, 12$.

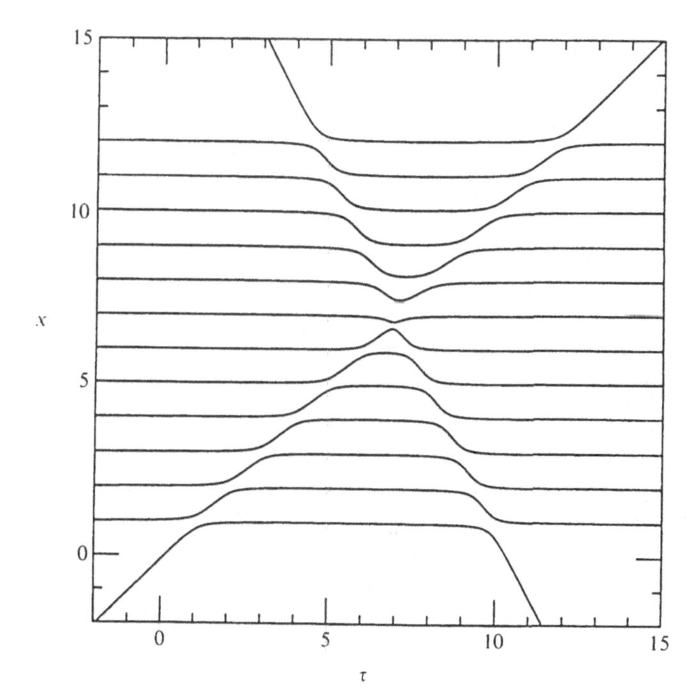

distribution in the small spacing region; fig. 5.6(*c*) has a long tail in the large curvature region. Although the statistical mechanics of solitons in a gCM system is anticipated to explain the above features, we have not as yet succeeded in its construction. Instead, we confine our discussion in the following to the statistical mechanics of the gCM system itself. The description below follows work by Gaspard *et al.* (1990).

Despite the complete integrability of the finite gCM system, statistical information is available from any finite interval in the infinite gCM system where dynamical mixing becomes possible. (Infinite dynamical systems

Fig. 5.4 Energy spectrum (E versus B) for odd-parity states of $m = 0$ Landau levels of lithium in an applied magnetic field B. Reduced-term-value plots for the outlined regions are given in the right-hand panels. (Courtesy of C. Iu *et al.*)

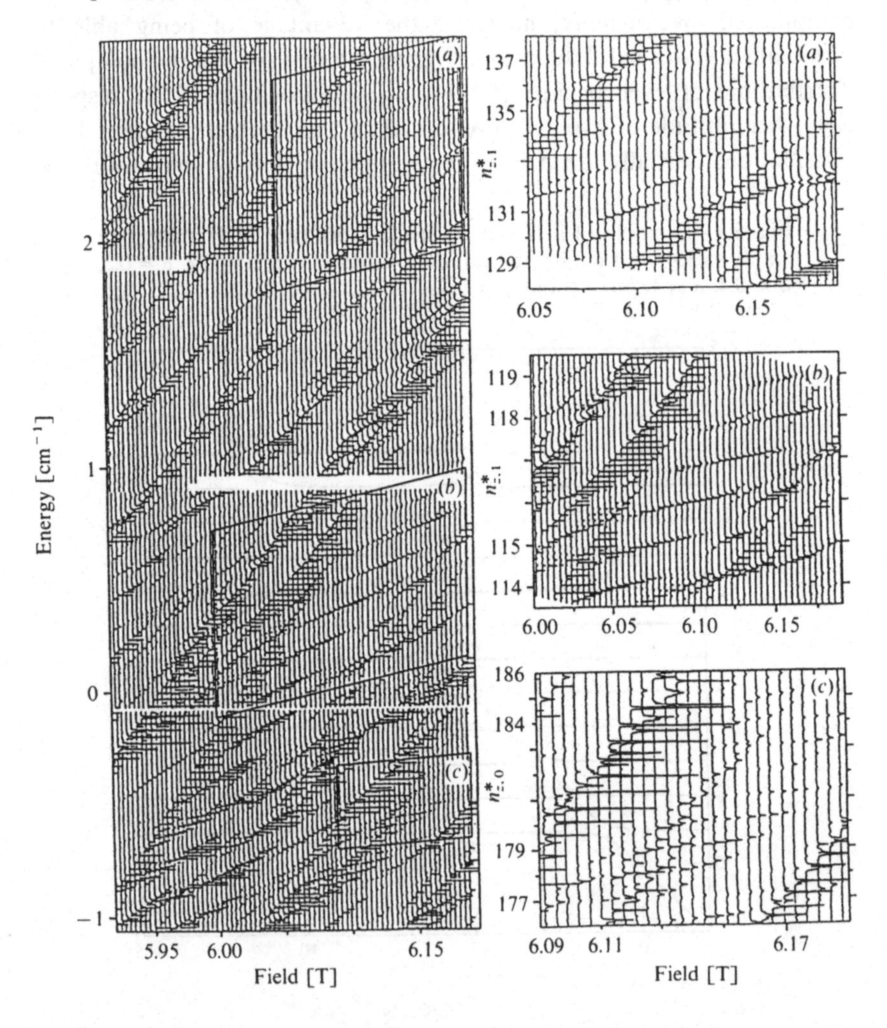

Fig. 5.5 Distribution of level spacings for the same system as in fig. 5.2 but with spin magnitude $S = 160$: (*a*) $B = 0$; (*b*) $B = 0.5$; (*c*) $B = 1$. Insets: magnified histograms but with fivefold finer cells.

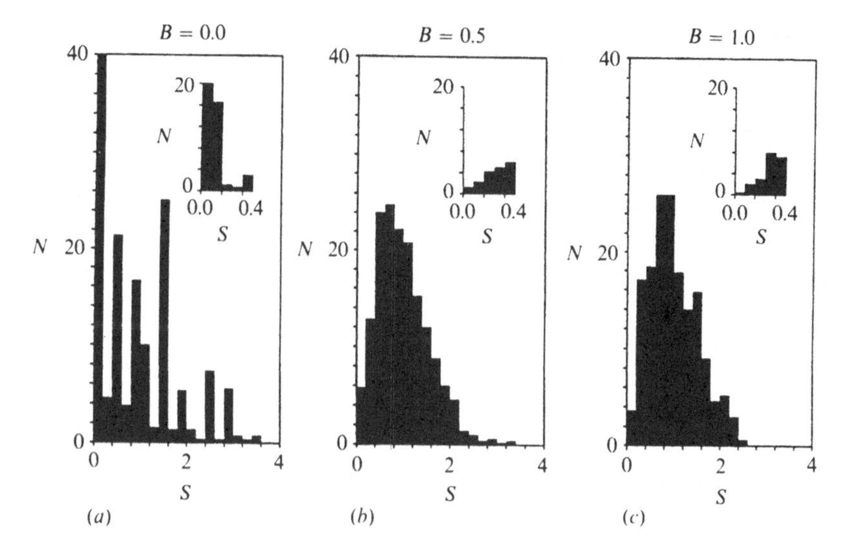

Fig. 5.6 Distribution of level curvatures for the same system as in fig. 5.5: (*a*) $B = 0$; (*b*) $B = 0.5$; (*c*) $B = 1$. Note the scale difference in ordinates between (*a*) and (*b*), (*c*). Curvatures are computed by using the discrete form of (5.94) with ΔB (corresponding to $\Delta\tau$) being 10^{-4}.

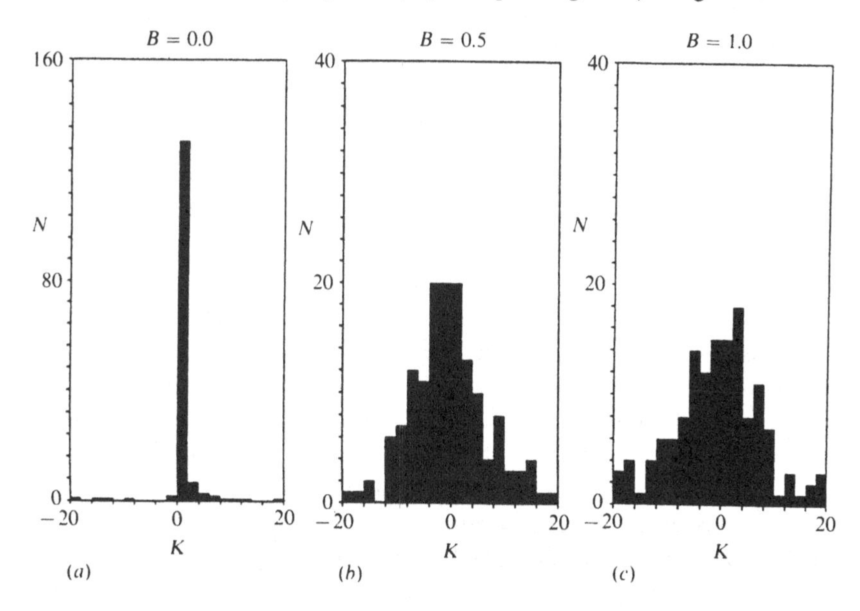

may have ergodic, mixing, or Bernoulli properties, as for an ideal gas or harmonic solid, while no local dynamical instability exists in these systems because of complete integrability of the corresponding finite dynamical system (Sinai, 1976; Sinai and Volkoviski, 1977; Goldstein *et al.*, 1975; Lanford and Lebowitz, 1975). We assume such results for infinite gCM systems.)

To study the gCM system, we need an invariant probability density defining a canonical ensemble. To simplify construction, we assume that the energy spectrum extends from $-\infty$ to $+\infty$ and is invariant under x-translations with a uniform density ρ. In actual quantum systems, the level density is usually non-uniform, but for the purpose of studying spectral fluctuations, uniform density is indispensable.

To construct the canonical ensemble, we introduce an intermediate canonical ensemble of systems with N particles in the interval $[-L/2, +L/2]$. At the endpoints, $-L/2$ and $+L/2$, we place hard walls where particles undergo elastic collisions. Between walls motion is ruled by the classical gCM Hamiltonian (5.11). Choosing dynamical variables $\{x_m\}$, $\{p_m\}$ and $\{\Lambda_{nm}\}$ with

$$\Lambda_{mn} \equiv \langle m|L^\dagger \cdot L|n\rangle \qquad \text{for } m \neq n, \tag{5.41}$$

we define a Gibbs measure:

$$\mathrm{d}M_{N,L} = \frac{1}{Z_{N,L}} \mathrm{e}^{-\beta \mathscr{H}_N} \prod_{1 \le m < n \le N} \prod_{a=0}^{\nu-1} \mathrm{d}x_m \, \mathrm{d}p_m \, \mathrm{d}\Lambda_{nm}^{(a)}, \tag{5.42}$$

where $\Lambda_{nm}^{(a)}$ are subvariables given in (5.30). The canonical ensemble will then be obtained in the limit where size L of the box increases indefinitely while keeping density ρ constant. The resulting measure depends on two parameters: density ρ and inverse temperature β. The numerical value of parameter β is determined from the variance of velocities $\{p_n(\tau)\}$ of the energy levels:

$$\langle(\Delta p_n)^2\rangle = \langle p_n^2\rangle - \langle p_n\rangle^2 = \frac{1}{\beta}. \tag{5.43}$$

Vanishing mean drift (i.e., $\langle p_n\rangle = 0$) is desirable in the observed spectrum. Otherwise, we should subtract this drift $\langle p_n\rangle$ from the motion of individual energy levels before comparing with predictions inferred from (5.42). To fulfill translational invariance of the canonical measure, we must now verify that a uniform density of energy levels is obtained for the canonical ensemble in the thermodynamics limit where L, $N \to \infty$ while N/L remains constant. This verification will eventually consolidate the thermodynamic formalism of random matrix theory.

5.5.1 Thermodynamic formalism of random matrix theory

Integration of the measure (5.42) over variables p_m and $\Lambda_{mn}^{(a)}$ yields the joint probability density of the energy levels:

$$f_{N,L}(x_1,\ldots,x_n) = \frac{1}{\mathcal{N}_{N,L}} \prod_{1\le i<j\le N} |x_i - x_j|^\nu \tag{5.44}$$

for $-L/2 \le x_i \le +L/2$. Equation (5.44) is nothing but the joint probability density encountered in random matrix theory, except for the weight functions. This fact implies the following: if the uniform level density is guaranteed, the canonical measure (5.42) for a gCM system is anticipated to yield known results from random matrix theory. The normalizing constant $\mathcal{N}_{N,L}$ in (5.44), level density, as well as spacing distribution can all be calculated starting from the generating functional defined by

$$I_N(u) = \int_{-1}^{1} \cdots \int_{-1}^{1} \prod_{k=1}^{N} u(x_k) \prod_{1\le i<j\le N} |x_i - x_j|^\nu \, dx_1 \cdots dx_N, \tag{5.45}$$

where we set the size of the box to $L = 2$ for simplicity. The normalizing constant is then given by

$$\mathcal{N}_{N,L} = I_N(1)(L/2)^{N+\nu N(N-1)/2}, \tag{5.46}$$

where $I_N(1)$ is the integral (5.45) with function $u(x) = 1$. Accordingly, the partition function becomes

$$Z_{N,L} = I_N(1)2^{N/2}[(\pi/\beta)^{1/2}L/2]^{N+\nu N(N-1)/2}. \tag{5.47}$$

In random matrix theory, Mehta and Gaudin (1960; 1961) developed methods to reduce integrals like (5.45) to the Fredholm determinant of an integral kernel composed of mutually orthogonal functions. Harmonic oscillator eigenfunctions are used in the case of Gaussian ensembles for which the joint probability density is (5.44) multiplied by the weight function $\exp(-1/2 \sum_k x_k^2)$. Several other ensembles, like Jacobi, Laguerre, or Legendre ensembles, have been considered by changing the weight function in order to use the corresponding orthogonal polynomials in reduction of the integral (5.45) by Mehta–Gaudin techniques. In this classification, (5.44) is the joint probability density of the Legendre ensembles because the weight function is 1 so that we need the Legendre polynomials to calculate (5.45). Within the Legendre ensembles (as well as within the other ones), exponent $\nu = 1$, 2 or 4 defines orthogonal, unitary, or symplectic ensembles, respectively. The procedure of analysing statistical mechanics of a gCM system is thus reduced to analysis of the Legendre ensemble.

We now derive the normalizing constant, level density and spacing distribution for the Legendre orthogonal ensemble with $\nu = 1$. This derivation will make readers understand that the spacing distribution for the Legendre orthogonal ensemble belongs to Dyson's universality class. While we follow closely Mehta's method, extension to the Legendre orthogonal ensemble is not straightforward. The additional difficulty is caused by nonvanishing of Legendre polynomials at boundaries of interval $[-1, 1]$, whereas the harmonic oscillator eigenfunctions used for Gaussian ensembles vanish at the boundaries of their domain of definition, namely $(-\infty, \infty)$. Choosing even values N for simplicity, we concentrate on calculation of the generating functional $I_N(u)$ in (5.45) with $\nu = 1$.

The domain $[-1, 1]^N$ is decomposed into $N!$ subdomains where the variables $\{x_i\}$ are ordered. Because the integrand of (5.45) is a symmetric function, $I_N(u)$ is equal to $N!$ times the integral of the same integrand but over any one of these subdomains. For definiteness, we choose

$$R = \{-1 < x_1 < x_2 < \cdots < x_N < 1\}. \tag{5.48}$$

Because the variables are now ordered, the product $\prod |x_i - x_j|$ is equal to the Vandermonde determinant

$$\begin{bmatrix} 1 & 1 & \dots & 1 \\ x_1 & x_2 & \dots & x_N \\ \vdots & \vdots & \vdots & \vdots \\ x_1^{N-1} & x_2^{N-1} & \dots & x_N^{N-1} \end{bmatrix}. \tag{5.49}$$

Using the property of Legendre polynomials that

$$P_n(x) = \frac{(2n)!}{2^n (n!)^2} x^n + O(x^{n-1}), \tag{5.50}$$

integral (5.45) becomes

$$I_n(u) = N! A_{N-1} \int \cdots \int_R \det [u(x_i) P_{j-1}(x_i)]_{i,j=1,\dots,N} \, dx_1 \cdots dx_N \tag{5.51}$$

with $A_{N-1} = \prod_{k=0}^{N-1} [2^k (k!)^2/(2k)!]$. Equation (5.51) can then be expressed as a Pfaffian

$$I_n(u) = N! A_{N-1} \{\det [f_{kl}]_{k,l=0,\dots,N-1}\}^{1/2} \tag{5.52}$$

with

$$f_{kl} = \int_{-1}^{1} dx \int_{-1}^{x} dy \, u(x) u(y) [P_k(x) P_l(y) - P_k(y) P_l(x)]. \tag{5.53}$$

Table 5.1 *Matrix elements used in calculation of $I_N(u)$.*

$\tilde{f}_{2i,2j}$	$\int_{-1}^{1} \mathrm{d}x \int_{-1}^{x} \mathrm{d}y\, u(x)u(y)[P_{2i}(x)P_{2j}(y) - P_{2i}(y)P_{2j}(x)]$
$\tilde{f}_{2i,2j-1}$	$\frac{1}{4j-1}\int_{-1}^{1} \mathrm{d}x \int_{-1}^{x} \mathrm{d}y\, u(x)u(y) \times [P_{2i}(x)P'_{2j}(y) - P_{2i}(y)P'_{2j}(x)]$
$\tilde{f}_{2i-1,2j-1}$	$\frac{1}{(4i-1)(4j-1)}\int_{-1}^{1} \mathrm{d}x \int_{-1}^{x} \mathrm{d}y\, u(x)u(y) \times [P'_{2i}(x)P'_{2j}(y) - P'_{2i}(y)P'_{2j}(x)]$
$\lambda_{ij}\ (=-\lambda_{ji})$	$\frac{1}{4}\int_{-1}^{1} \mathrm{d}x \int_{-1}^{x} \mathrm{d}y\, u(x)\, u(y)[P_{2i}(x)P_{2j}(y) - P_{2i}(y)P_{2j}(x)]$
g_{ij}	$\frac{4j+1}{4}\int_{-1}^{1} \mathrm{d}x \int_{-1}^{x} \mathrm{d}y\, u(x)u(y) \times [P_{2i}(x)\phi_{2j}(y) - P_{2i}(y)\phi_{2j}(x)]$
$\mu_{ij}\ (=-\mu_{ji})$	$\frac{(4i+1)(4j+1)}{4}\int_{-1}^{1} \mathrm{d}x \int_{-1}^{x} \mathrm{d}y\, u(x)u(y) \times [\phi_{2i}(x)\phi_{2j}(y) - \phi_{2i}(y)\phi_{2j}(x)]$

Note that $\phi_{2i}(x) \equiv P'_{2i}(x) - P'_N(x)$.

The indices k and l are separated into even and odd numbers by permutations of lines and columns in det $[f_{kl}]$ so that indices are ordered like $k, l = 0, 2, 4, \ldots, N-2;\ N-1, 1, 3, \ldots, N-3$. Note that the odd integer $N-1$ is placed in front of the other odd integers in order to deal with nonvanishing of Legendre polynomials at ± 1, as below.

Using the following property of Legendre polynomials,

$$\begin{aligned} P'_{2j}(x) &\equiv \mathrm{d}P_{2j}(x)/\mathrm{d}x \\ &= (4j-1)P_{2j-1}(x) + (4j-5)P_{2j-3}(x) + \cdots + 3P_1(x) \end{aligned} \tag{5.54}$$

and forming linear combinations of lines and columns, det $[f_{kl}]$ is transformed into det $[\tilde{f}_{kl}]$ with elements given in table 5.1. Columns $\tilde{f}_{kl}$ with $l = 1, 3, \ldots, N-3$ are replaced by

$$(2l+1)(2l+3)\left(\tilde{f}_{kl} - \frac{2N-1}{2l+1}\tilde{f}_{k,N-1}\right). \tag{5.55}$$

The column $\tilde{f}_{k,N-1}$ is replaced by $-\tilde{f}_{k,N-1}$. Similar transformations are carried out on the lines $k = N-1, 1, 3, \ldots, N-3$. All the lines are then divided by 4. Accordingly,

$$\det[\tilde{f}_{kl}] = \left[\frac{2^N}{3\cdot 5\cdot 7\cdots(2N-3)(2N-1)}\right]^2 \det\begin{bmatrix}\lambda_{ij} & g_{ij}\\ -g_{ji} & \mu_{ij}\end{bmatrix}, \tag{5.56}$$

where λ_{ij}, g_{ij}, etc. are given in table 5.1 and $i, j = 0, 1, 2, \ldots, N/2 - 1$. The row and column labeled by k or $l = N-1$ in $\tilde{f}_{kl}$ are transformed into the column $j = 0$ of g_{ij} and μ_{ij} as well as the row $i = 0$ of μ_{ij}. They are no longer separated from other columns and lines in the general notation of g_{ij} and μ_{ij} in table 5.1. In this way we arrive at the result:

$$I_N(u) = 2^N A_N \left\{\det\begin{bmatrix}\lambda_{ij} & g_{ij}\\ -g_{ji} & \mu_{ij}\end{bmatrix}\right\}^{1/2}. \tag{5.57}$$

Equation (5.57) can result in various results from random matrix theory. Using $u(x) = 1$ in table 5.1, we obtain quite simple values:

$$\lambda_{ij} = \mu_{ij} = 0, \qquad g_{ij} = \delta_{ij}. \tag{5.58}$$

The normalizing constant is thereby

$$\mathcal{N}_{N,2} = I_N(1) = 2^N A_N, \tag{5.59}$$

so that the partition function for orthogonal systems ($\nu = 1$) behaves like

$$Z_{N,L} \simeq 2^{3N/2}\cdot[(\pi/\beta)^{1/2} L/4]^{N(N+1)/2} \tag{5.60}$$

for $N \to \infty$. (Note that $A_N \to 1$ in this limit.)

The level density is obtained from the functional derivatives with respect to function $a(x)$ of functional (5.57) where function $u(x)$ is replaced by $1 + a(x)$ (Dyson, 1962a–1962d). To be explicit,

$$\sigma_{N,2}(x) \equiv \left[\frac{\delta}{\delta a(x)}\frac{I_N(1+a)}{I_N(1)}\right]_{a=0} = \left[\frac{\delta}{\delta a(x)}\sum_{i=0}^{N/2-1} v_{ii}\right]_{a=0}, \tag{5.61}$$

where $v_{ij} \equiv g_{ij} - \delta_{ij}$. Using expansion (5.54) of even Legendre polynomial derivatives together with its odd polynomial version, we obtain

$$\sigma_{N,2}(x) = \sum_{k=0}^{N-1}\frac{2k+1}{2}[P_k(x)]^2 - \tfrac{1}{2}P_N(x)P'_{N-1}(x). \tag{5.62}$$

Using the Christoffel–Darboux formula and the derivative recurrence formula, we finally have

$$\sigma_{N,2}(x) = \frac{N}{2(1-x^2)}[NP_{N-1}^2(x) + (N+1)P_N^2(x) - (2N+1)xP_{N-1}(x)P_N(x)]. \qquad (5.63)$$

Using large-order expansion of Legendre polynomials together with rescaling $x \to 2x/L$, level density near $x = 0$ is given by

$$\sigma_{N,L}(x) \simeq \frac{2N}{\pi L}[1 - (2x/L)^2]^{-1/2} \qquad (5.64)$$

for $N \to \infty$. On the other hand, using in (5.62) the odd version of (5.54) and $P_n(\pm 1) = (\pm 1)^n$, $\sigma_{N,L}(x)$ is found to increase to

$$\sigma_{N,L}(\pm L/2) = \frac{N(N+1)}{2L}$$

near the walls of the box. Inside the box and far from the walls, however, level density (5.64) is linear in N in contrast to Wigner's 'semicircle law' for Gaussian ensembles given by

$$\sigma_N(x) = \frac{1}{\pi}(2N - x^2)^{1/2}. \qquad (5.65)$$

Whilst (5.65) has the disadvantage of increasing like $N^{1/2}$, density (5.64) naturally reaches a uniform value

$$\rho \equiv \frac{2N}{\pi L} \qquad (5.66)$$

in the thermodynamic limit $N, L \to \infty$ with N/L a constant. So, the Gibbs measure in (5.42) has been found to construct the canonical ensemble in this limit. To conclude, the Legendre ensemble emerging from the gCM system ensures most naturally the thermodynamic formalism of random-matrix theory.

5.5.2 *Level spacing distribution*

The level spacing distribution is one of the major results of random matrix theory. Quantum-mechanical treatment of fully chaotic systems has been addressed to exhibit a 'universal' level spacing distribution (e.g., McDonald and Kaufman, 1979; Bohigas *et al.*, 1984). In this subsection we shall argue how the generating functional (5.57) leads to a universal level spacing distribution.

The dimensionless spacing between two nearest-neighbor levels $x_x < x_{n+1}$ is defined by

$$S = (\text{spacing})/(\text{mean spacing}) = \rho(x_{n+1} - x_n). \tag{5.67}$$

The spacing density $p(S)$ is the second derivative of the probability $\varepsilon(S)$ that an interval of size S is empty of eigenvalues

$$p(S) = \frac{d^2\varepsilon(S)}{dS^2}. \tag{5.68}$$

$\varepsilon(S)$ itself is available by the limiting behavior of probability $\varepsilon_N(\theta)$ that interval $[-\theta, +\theta]$ with $0 < \theta < 1$ is empty of levels (Mehta, 1967). The latter probability is given by

$$\varepsilon_N(\theta) = I_N(u)/I_N(1) \tag{5.69}$$

with

$$u(x) = \begin{cases} 0 & -\theta < x < +\theta, \\ 1 & \theta < |x|. \end{cases} \tag{5.70}$$

Since (5.70) is an even function of x, we have $\lambda_{ij} = \mu_{ij} = 0$, and thereby

$$\varepsilon_N(\theta) = \det [g_{ij}]_{i,j=0,\ldots,N/2-1}. \tag{5.71}$$

After integration, we get

$$g_{ij} = \delta_{ij} - M_{ij} \tag{5.72}$$

with

$$M_{ij}(\theta) = \frac{4j+1}{2} \int_{-\theta}^{\theta} P_{2i}(x)[P_{2j}(x) - P_N(x)]\, dx. \tag{5.73}$$

We now take a subtle method to calculate the kernel function (Mehta, 1967). Let us note the identity

$$\det [\hat{I} - \hat{M}(\theta)] = \prod_{k=0}^{N/2-1} (1 - \lambda_k), \tag{5.74}$$

where $\{\lambda_k\}$ are the eigenvalues of the matrix (5.73). They are also the eigenvalues,

$$\lambda f(x) = \int_{-\theta}^{\theta} \mathscr{K}_N(x, y) f(y)\, dy \tag{5.75}$$

of the kernel

$$\mathscr{K}_N(x, y) = \sum_{i=0}^{N/2-1} \frac{4i+1}{2} P_{2i}(x)[P_{2i}(y) - P_N(y)], \tag{5.76}$$

acting on the functions

$$f(x) = \sum_{i=0}^{N/2-1} c_i P_{2i}(x). \qquad (5.77)$$

We note that kernel (5.76) characterizes the Legendre orthogonal ensemble and differs from kernels calculated previously for other ensembles.

In the limit $N \to \infty$, using the Christoffel-Darboux formula together with large-order asymptotic expansion of Legendre polynomials

$$P_{2m}\left(\frac{\pi\xi}{2m}\right) \simeq \frac{(-1)^m}{(\pi m)^{1/2}} \cos \pi\xi, \qquad (5.78a)$$

and

$$P_{2m+1}\left(\frac{\pi\xi}{2m}\right) \simeq \frac{(-1)^m}{(\pi m)^{1/2}} \sin \pi\xi, \qquad (5.78b)$$

the kernel in (5.76) becomes the universal one (Mehta, 1967):

$$\lim_{N\to\infty} \frac{\pi}{N} \mathscr{K}_N\left(\frac{\pi\xi}{N}, \frac{\pi\eta}{N}\right) = \frac{1}{2}\left[\frac{\sin \pi(\xi - \eta)}{\pi(\xi - \eta)} + \frac{\sin \pi(\xi + \eta)}{\pi(\xi + \eta)}\right]. \qquad (5.79)$$

Accordingly,

$$\varepsilon(S) \equiv \lim_{N\to\infty} \varepsilon_N(\pi S/2N) = \prod_{i=0}^{\infty} [1 - S\gamma_{2i}^2(S)], \qquad (5.80a)$$

where

$$\gamma_{2i}(S) = \frac{1}{2f_{2i}(0; S)} \int_{-1}^{1} f_{2i}(z; S)\, \mathrm{d}z \qquad (5.80b)$$

in terms of the S-dependent spheroidal functions $f_0, f_2, f_4, \ldots$.

The result in (5.80) is identical to the universal function obtained by Gaudin, Mehta and Dyson for Gaussian and circular ensembles (Mehta and Gaudin, 1960, 1961; Mehta and Dyson, 1963; Mehta, 1967). From (5.68), we have

$$p_{\mathrm{OE}}(S) = \frac{\pi^2}{6} S - \frac{\pi^4}{60} S^3 + \frac{\pi^4}{270} S^4 + \frac{\pi^6}{1680} S^5 + \cdots. \qquad (5.81a)$$

This result completes the proof by Leff (1964) that the spacing distribution for a Legendre unitary ensemble is the corresponding universal spacing

$$p_{\mathrm{UE}}(S) = \frac{\pi^2}{3} S^2 - \frac{2\pi^4}{45} S^4 + \frac{\pi^6}{315} S^6 + \cdots \qquad (5.81b)$$

with quadratic repulsion. Assuming universality of different ensembles, we conjecture that the spacing density for a Legendre symplectic ensemble will be the Mehta–Dyson one (Mehta and Dyson, 1963):

$$p_{\mathrm{SE}}(S) = \frac{2^4\pi^4}{135}S^4 - \frac{2^7\pi^6}{4725}S^6 + \cdots \tag{5.81c}$$

with quartic repulsion.

The statistical mechanics of gCM systems has thus succeeded in explaining why level spacing distribution for classically chaotic systems should obey the same universal distributions (5.81) as those of random matrix theory.

5.5.3 Fate of Brownian motion model

It is worthwhile to comment here on Dyson's Brownian motion model for eigenvalues of random matrices. Dyson (1962a–1962d; 1972) was the first to conceive the idea of deriving joint probability density as (5.44) from a level dynamics standpoint. He remarked (Dyson, 1962c), 'After considerable and fruitless efforts to develop a Newtonian theory of ensembles, we discovered that the correct procedure is quite different... The x_j (eigenvalues) should be interpreted as positions of particles in Brownian motion. This means that the particles do not have well-defined velocities, nor do they possess inertia'. So long as he confined his argument to random matrices, his remark was right. His theory of Brownian motion in the form of stochastic differential equations is sketched below.

For a Gaussian orthogonal ensemble, the stochastic differential equation for a τ-dependent (τ is 'quasitime' as in previous sections) real symmetric matrix $H(\tau)$ is given by

$$\mathrm{d}H(\tau) = -H\,\mathrm{d}\tau + \mathrm{d}V(\tau), \tag{5.82}$$

where $V(\tau)$ is also a member of the Gaussian orthogonal ensemble and $\mathrm{d}V(\tau)$ denotes a fundamental Brownian step obeying Wiener processes

$$\overline{\mathrm{d}V_{nm}} = 0, \tag{5.83a}$$

$$\overline{\mathrm{d}V_{nm}\,\mathrm{d}V_{n'm'}} = (\delta_{nn'}\delta_{mm'} + \delta_{nm'}\delta_{mn'})\xi^2\,\mathrm{d}\tau. \tag{5.83b}$$

The essential point is that $\mathrm{d}V$ is of the order of $(\mathrm{d}\tau)^{1/2}$. As usual, let us examine the eigenvalue problem,

$$H(\tau)|n(\tau)\rangle = x_n(\tau)|n(\tau)\rangle. \tag{5.84}$$

Replacing τ in (5.84) by $\tau + \mathrm{d}\tau$ and using expansions with respect to $(\mathrm{d}\tau)^{1/2}$

$$H(\tau + \mathrm{d}\tau) = H(\tau) + \mathrm{d}H(\tau), \tag{5.85a}$$

$$x_n(\tau + \mathrm{d}\tau) = x_n(\tau) + \mathrm{d}x_n(\tau), \tag{5.85b}$$

$$|n(\tau + \mathrm{d}\tau)\rangle = \sum_m c_{nm}(\tau + \mathrm{d}\tau)|m(\tau)\rangle \tag{5.85c}$$

with

$$\mathrm{d}x_n = \mathrm{d}x_n^{(1/2)} + \mathrm{d}x_n^{(1)} + \cdots, \tag{5.86a}$$

$$c_{nm} = c_{nm}^{(0)} + c_{nm}^{(1/2)} + c_{nm}^{(1)} + \cdots, \tag{5.86b}$$

standard perturbation theory (up to order $\mathrm{d}\tau$) yields

$$\mathrm{d}x_n(\tau) = -x_n\,\mathrm{d}\tau + \sum_{m(\neq n)} \frac{\xi^2}{x_n - x_m}\,\mathrm{d}\tau + \mathrm{d}W_{nn}(\tau) \tag{5.87a}$$

and

$$\mathrm{d}|n(\tau)\rangle = -\frac{1}{2}\sum_{m(\neq n)} \frac{\xi^2}{(x_n - x_m)^2}|n(\tau)\rangle\,\mathrm{d}\tau + \sum_{m(\neq n)} \frac{\mathrm{d}W_{mn}}{x_n - x_m}|m(\tau)\rangle \tag{5.87b}$$

with $\mathrm{d}W$ defined by

$$\overline{\mathrm{d}W_{nm}} = 0, \tag{5.88a}$$

$$\overline{\mathrm{d}W_{nm}(\tau)\,\mathrm{d}W_{n'm'}(\tau)} = (\delta_{nn'}\delta_{mm'} + \delta_{nm'}\delta_{mn'})\xi^2\,\mathrm{d}\tau. \tag{5.88b}$$

We find that the equation for eigenvalues, (5.87*a*), which is decoupled from eigenstates, describes the one-dimensional Brownian motion of particles of unit charge and unit mass subject to both on-site harmonic potential and repulsive two-dimensional Coulomb potential; ξ^{-2} plays the role of friction coefficient which fixes diffusion rate.

Equation (5.87*a*) with (5.88) is identical to the Smoluchowski equation for time-dependent probability density $P(\{x_n\};\tau)$:

$$\xi^{-2}\frac{\partial}{\partial\tau}P = \sum_n \frac{\partial}{\partial x_n}\left\{\frac{\partial}{\partial x_n} - F_n\right\}P, \tag{5.89}$$

where $F_n(= -\partial U/\partial x_n)$ is a systematic force derived from the effective potential,

$$U = (2\xi^2)^{-1}\sum_n x_n^2 - \sum_{m(\neq n)} \ln|x_n - x_m|. \tag{5.90}$$

The stationary ($\tau \to \infty$) solution of (5.89) yields the joint probability density for a Gaussian orthogonal ensemble:

$$P(\{x_n\}; \tau \to \infty) = \prod_{n<m} |x_n - x_m| \exp\left(-\sum_n x_n^2/(2\xi^2) \right), \tag{5.91}$$

which, as noted in section 5.5.1, belongs to the same universality class as Legendre orthogonal ensembles. This completes Dyson's scenario. Similar arguments apply to Gaussian unitary and symplectic ensembles.

Let us note further that (5.87*a*) with (5.88) is the overdamped limit ($\ddot{x}_n \simeq 0$) of Langevin equations given by

$$\ddot{x}_n = -\xi^{-2}\dot{x}_n + F_n + w_n(\tau) \tag{5.92}$$

with

$$\overline{w_n(\tau)} = 0, \tag{5.93a}$$

$$\overline{w_n(\tau)w_m(\tau')} = 2\delta_{nm}\delta(\tau - \tau'). \tag{5.93b}$$

Here $w_n(\tau)$ represents Gaussian white noise. Despite some similarity between the gCM system in (5.13) with (5.11) and Dyson's Brownian motion model, there are clear differences between these two systems. The gCM system describes a conservative Newtonian motion for particles subject to an inter-particle interaction potential of power-law type. The equations for particle positions are strongly coupled with their internal degree of freedom, i.e., eigenstates. In contrast, Dyson's Brownian motion model in (5.87) with (5.88) deals with the overdamped limit of a Langevin equation for particles with both on-site harmonic potential and inter-particle interaction potential of a logarithmic Coulomb type. A crucial point is that no coupling between particle positions and internal degree of freedom exists in this case.

The distinct contrasts addressed above are attributed to the nature of the originating models. The Brownian motion model originates in random matrices whereas the gCM system originates in deterministic Hamiltonians. It is astonishing that such utterly different starting models have led to joint probability distributions of the same universality class. Being concerned with deterministic quantum chaos, however, we are absolutely free of Brownian motion models for random matrices and ought to be involved largely in managing conservative Newtonian systems (gCM).

5.5.4 Curvature distribution: beyond random matrix theory

The traditional framework of random matrix theory reveals only statistical information when a set of nonintegrability parameters are fixed. The statistical mechanics of a gCM system, on the other hand, can also predict distribution functions for parameter-dependent quantities.

Among them, the most important quantity is level curvature given by

$$K_i \equiv \lim_{\Delta\tau\to 0} \frac{\Delta^2 x_i}{\Delta\tau^2} = \ddot{x}_i(\tau). \tag{5.94}$$

Curvature in (5.94) takes large values for a pair of adjacent levels at avoided crossings and constitutes a promising indicator of quantum chaos. The role of curvatures in energy spectra was originally pointed out by Pomphrey (1974). In practice, numerical computation of curvatures K_i can be done by using the discrete form of the second-order derivative in (5.94) with a small but finite $\Delta\tau$. (If we choose a value $\Delta\tau$ larger than the effective width of avoided crossings, anomalously large values for K_i may be available. This artificially induced problem can be overcome by taking a much smaller value $\Delta\tau$.) In chapter 3 we have already examined statistical behaviors of curvatures, indicating remarkable scaling behaviors of fluctuations of curvatures in the fully chaotic region. In (5.94) $\ddot{x}_i(\tau)$ is nothing but the acceleration of the ith particle in gCM systems, so that

$$\ddot{x}_i(\tau) = -\frac{\partial \mathscr{H}_N}{\partial x_i} \tag{5.95}$$

(see (5.13)). Without loss of generality, we choose x_1 as a sample level and derive the probability density for curvature $\ddot{x}_1$ to take the value K:

$$P(K) \equiv \lim_{\substack{N,L\to\infty \\ N/L=\text{const}}} P_{N,L}(K) \tag{5.96}$$

with

$$P_{N,L}(K) \equiv \int \delta\left(K + \frac{\partial \mathscr{H}_N}{\partial x_1}\right) \mathrm{d}M_{N,L}, \tag{5.97}$$

where $\mathrm{d}M_{N,L}$ is the Gibbs measure (5.42). (The derivation given below was also developed by Gaspard *et al.* (1990).) From (5.13), the equation $K + (\partial \mathscr{H}_N/\partial x_1) = 0$ possesses $N - 1$ roots $\{x_1^{(k)}\}$ for any value of K. These zeros can be evaluated when K is large and positive, because $\partial \mathscr{H}_N/\partial x_1$ is a function of x_1 which diverges in the vicinity of levels different from x_1 (see fig. 5.7). Choosing K positive, we obtain

$$x_1^{(k)} \simeq x_k + \left[\frac{2}{K}\sum_a (\Lambda_{1k}^{(a)})^2\right]^{1/3}. \tag{5.98}$$

Consequently, we get

$$P_{N,L}(K) = \sum_{k=2}^{N} \int \left| \frac{\partial^2 \mathscr{H}_N}{\partial x_1^2} \right|^{-1} \delta[x_1 - x_1^{(k)}(K)] \, \mathrm{d}M_{N,L} \tag{5.99}$$

with

$$\frac{\partial^2 \mathscr{H}_N}{\partial x_1^2}(x_1^{(k)}) \simeq 3\left[\frac{2}{K^4}\sum_a (\Lambda_{1k}^{(a)})^2\right]^{-1/3}. \tag{5.100}$$

Equation (5.99) is evaluated asymptotically for large positive values of K. After integration over the variables x_1, $\{p_n\}$ and $\{\Lambda_{mn}^{(a)}\}$ with m and n different from 1, we obtain

$$P_{N,L}(K) = \frac{1}{Z_{N,L}}\left(\frac{2\pi}{\beta}\right)^{N/2} \sum_{k=2}^{N} \int \left| \frac{\partial^2 \mathscr{H}_N}{\partial x_1^2}(x_1^{(k)}) \right|^{-1} \exp\left[-\beta \sum_{n=2}^{N} \sum_a \frac{(\Lambda_{1n}^{(a)})^2}{(x_1^{(k)} - x_n)^2}\right]$$
$$\times \prod_{2 \le i < j \le N} [(\pi/\beta)^{1/2}|x_i - x_j|]^{\nu} \, \mathrm{d}x_2 \cdots \mathrm{d}x_N \prod_a \mathrm{d}\Lambda_{12}^{(a)} \cdots \Lambda_{1N}^{(a)}. \tag{5.101}$$

From the asymptotic expression (5.98) for root $x_1^{(k)}$, we have

$$\sum_{n=2}^{N} \sum_a \frac{(\Lambda_{1n}^{(a)})^2}{(x_1^{(k)} - x_n)^2} = \left[\frac{K^2}{4}\sum_a (\Lambda_{1k}^{(a)})^2\right]^{1/3} + \sum_{\substack{n=2 \\ n \neq k}}^{N} \sum_a \frac{(\Lambda_{1n}^{(a)})^2}{(x_k - x_n)^2} + O(K^{-1/3}). \tag{5.102}$$

Replacing the second derivative of the gCM Hamiltonian by (5.100) and integrating over variable $\{\Lambda_{1n}^{(a)}\}$ with $n \neq k$ in the kth term, (5.101)

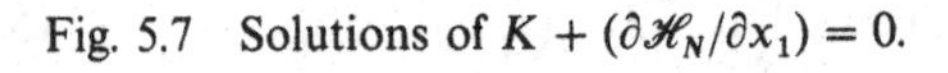
Fig. 5.7 Solutions of $K + (\partial \mathscr{H}_N / \partial x_1) = 0$.

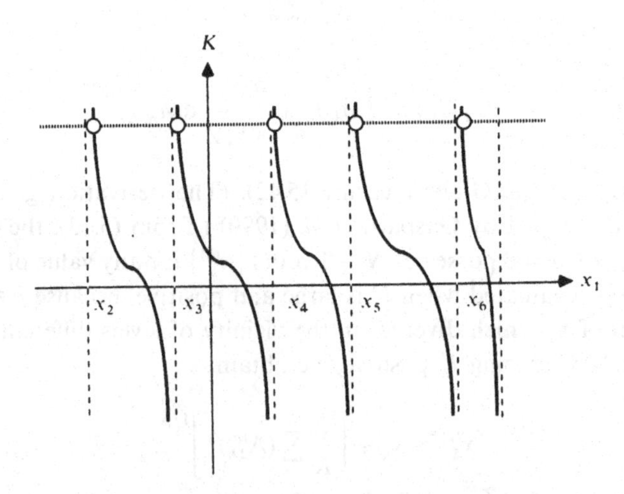

becomes

$$P_{N,L}(K) \simeq \frac{1}{Z_{N,L}}\left(\frac{2\pi}{\beta}\right)^{N/2} \sum_{k=2}^{N} \mathscr{I}_{\nu}^{(k)}$$

$$\times \int_{[-L/2,L/2]^{N-1}} \prod_{2 \le i < j \le N} [(\pi/\beta)^{1/2}|x_i - x_j|]^{\nu}$$

$$\times \prod_{\substack{n=2 \\ n \neq k}} [(\pi/\beta)^{1/2}|x_k - x_n|]^{\nu} \, dx_2 \cdots dx_{\nu}, \tag{5.103}$$

where

$$\mathscr{I}_{\nu}^{(k)} = \int_{(-\infty,+\infty)^{\nu}} \frac{1}{3}\left[\frac{2}{K^4}\sum_a (\Lambda_{1k}^{(a)})^2\right]^{1/3}$$

$$\times \exp\left\{-\beta\left[\frac{K^2}{4}\sum_a (\Lambda_{1k}^{(a)})^2\right]^{1/3}\right\} \prod_{a=0}^{\nu-1} d\Lambda_{1k}^{(a)}, \tag{5.104}$$

with $0 \le a \le \nu - 1$. This integral is independent of index k and is given by

$$\mathscr{I}_{\nu}^{(k)} = \frac{\pi^{\nu/2} 2^{2\nu-3}(\nu+2)!}{\beta^{3\nu/2+1} K^{\nu+2}}. \tag{5.105}$$

Finally, introducing

$$\mathscr{M}_{N,L} \equiv \sum_{k=2}^{N} \int_{-L/2}^{+L/2} \cdots \int_{-L/2}^{+L/2} \prod_{2 \le i < j \le N} |x_i - x_j|^{\nu} \prod_{\substack{n=2 \\ n \neq k}}^{N} |x_k - x_n|^{\nu} \, dx_2 \cdots dx_N, \tag{5.106}$$

and using (5.46), (5.47), we obtain

$$P_{N,L}(K) \simeq \frac{\mathscr{M}_{N,L}}{\mathscr{N}_{N,L}} \frac{2^{2\nu-3}(\nu+2)!}{\beta^{\nu+1} K^{\nu+2}}, \tag{5.107}$$

where $\mathscr{N}_{N,L}$ is the normalizing constant (5.46). The ratio $\mathscr{M}_{N,L}/\mathscr{N}_{N,L}$ in (5.107) can be rewritten as the following mean value over the joint probablity density (5.44)

$$\frac{\mathscr{M}_{N,L}}{\mathscr{N}_{N,L}} = \left\langle \sum_{k(\neq 1)} \frac{\delta(x_1 - x_k)}{|x_1 - x_k|^{\nu}} \right\rangle. \tag{5.108}$$

In the thermodynamic limit and far from the hard walls, this expression is related to spacing density at zero spacing.

Evaluation of the mean value (5.108) for a uniform spectrum of density

ρ is quite easy. Let x_0 be any eigenvalue in a sequence,

$$\cdots < x_{-2} < x_{-1} < x_0 < x_1 < x_2 < \cdots, \tag{5.109}$$

of ordered eigenvalues. Distribution $\delta(x)$ in (5.108) is replaced by the function,

$$\Delta_\varepsilon(x) = \begin{cases} 0 & \text{for } |x| > \varepsilon/2, \\ 1/\varepsilon & \text{for } |x| < \varepsilon/2. \end{cases} \tag{5.110}$$

Then (5.108) becomes a series of terms each containing a pair of mean values:

$$\sum_{m=0}^{\infty} \left\{ \left\langle \frac{\Delta_\varepsilon(x_0 - x_{m+1})}{|x_0 - x_{m+1}|^\nu} \right\rangle + \left\langle \frac{\Delta_\varepsilon(x_0 - x_{-m-1})}{|x_0 - x_{-m-1}|^\nu} \right\rangle \right\}. \tag{5.111}$$

The mth term is given by

$$\rho^{\nu+1} \frac{2}{\varepsilon} \int_0^{\varepsilon/2} S^{-\nu} p_\nu^{(m)}(S)\, \mathrm{d}S \qquad (m = 0, 1, 2, \ldots), \tag{5.112}$$

where $p_\nu^{(m)}(S)$ is the density of the mth order spacing distribution. For $m = 0$, $p_\nu^{(0)}(S)$ is the nearest-neighbor spacing density given by (5.81*a*)–(5.81*c*) for $\nu = 1, 2$ and 4 respectively. Noting that $p_\nu^{(m)}(S) \simeq S^\alpha$ with $\alpha > \nu$ for $m \geq 1$, (5.112) remains nonvanishing only for $m = 1$ in the limit $\varepsilon \to 0$. In this limit, therefore, (5.111) becomes

$$\lim_{\substack{N,L \to \infty \\ N/L = \text{const}}} \frac{\mathscr{M}_{N,L}}{\mathscr{N}_{N,L}} = \rho^{\nu+1} \lim_{S \to 0} S^{-\nu} p_\nu(S) \equiv D_\nu \rho^{\nu+1}, \tag{5.113}$$

where $p_\nu(S) \equiv p_\nu^{(0)}(S)$ and the constants D_ν are $\pi^2/6$, $\pi^2/3$ and $2^4\pi^4/135$ for $\nu = 1$, 2 and 4, respectively (see (5.81)). A similar result is available for $K < 0$.

Finally, the tail of the curvature distribution is given for each ensemble by

$$P_{\mathrm{OE}}(K) = \frac{\pi^2}{2} \left(\frac{\rho}{\beta}\right)^2 \frac{1}{|K|^3} + \cdots, \tag{5.114a}$$

$$P_{\mathrm{UE}}(K) = 2^4 \pi^2 \left(\frac{\rho}{\beta}\right)^3 \frac{1}{K^4} + \cdots, \tag{5.114b}$$

$$P_{\mathrm{SE}}(K) = \frac{2^{13}\pi^4}{3} \left(\frac{\rho}{\beta}\right)^5 \frac{1}{K^6} + \cdots, \tag{5.114c}$$

for large $|K|$. We note that the powers of parameters β and ρ are consistent with the fact that $\beta K/\rho$ is the dimensionless curvature. This fact also indicates that scaling of curvature is needed in the presence of a non-uniform spectrum. The curvature distributions (5.114) are normalizable so that they have a probabilistic interpretation. Their mean value is zero because densities are symmetric under $K \to -K$. The variance is infinite for the orthogonal ensemble but finite for unitary and symplectic ensembles. The former outcome is in conformity with the numerical result for quantum spin systems (Nakamura *et al.*, 1985) described in chapter 3.

In order to check the asymptotic expressions (5.114) of curvature densities, parametric motion of eigenvalues was studied by Takami and Hasegawa (1992) for deterministic Hamiltonian systems. They chose two prototypes of quantum chaos, i.e., a stadium billiard and a kicked rotator, to see both universality and nonuniversality aspects of curvature distributions. Figure 5.8 shows the τ-dependent energy spectra. Here τ prescribes the ratio of the semicircle radius over a half-length of straight line segment for stadium billiards and kicked strength for kicked rotators. Figure 5.8(*a*) involves remarkable soliton-like structures embedded in a sea of irregular spectra. The presence of such regular structures proves false our simple understanding of the quantum analog of fully chaotic K systems. The soliton-like structures emanate from (unstable) periodic bouncing, for instance, between a pair of parallel flat walls of the stadium.

Computed curvature distributions $P(K)$ are given (in logarithmic

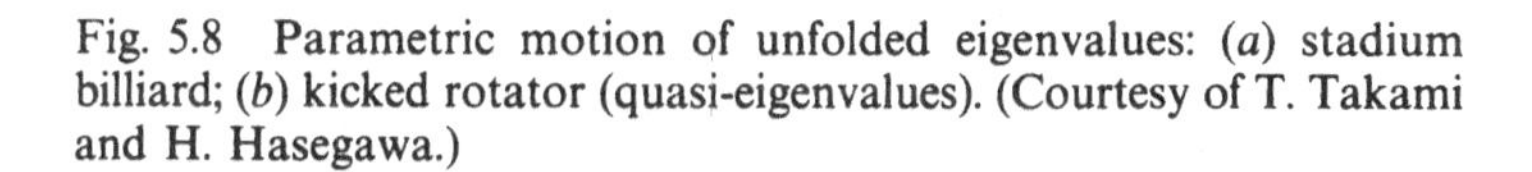

Fig. 5.8 Parametric motion of unfolded eigenvalues: (*a*) stadium billiard; (*b*) kicked rotator (quasi-eigenvalues). (Courtesy of T. Takami and H. Hasegawa.)

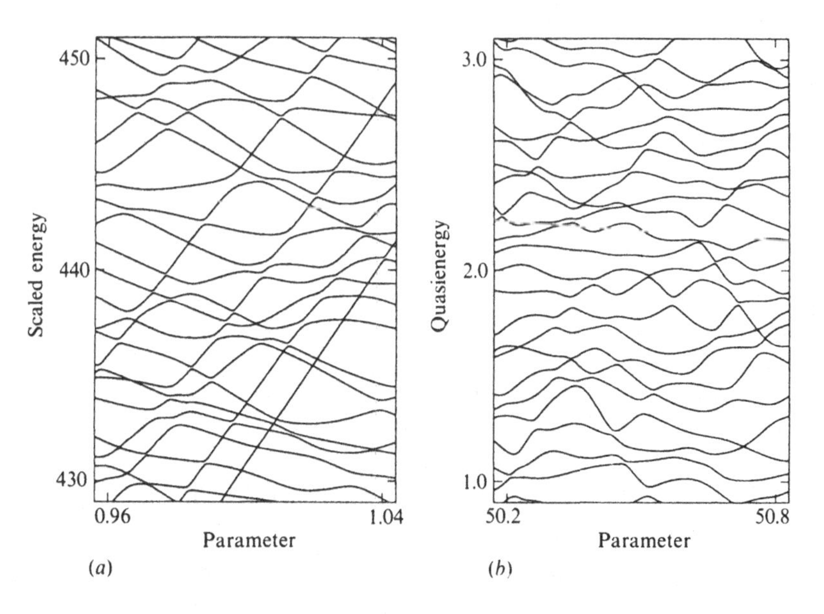

scales) in fig. 5.9. The tails of $P(K)$ for both stadium billiards and kicked rotator are found to obey a $|K|^{-3}$ law in agreement with theoretical prediction for $P_{OE}(K)$ in (5.114). (At the same time, nearest-neighbor spacing distributions show a universal linear S-dependence near $S = 0$.) Stadium billiards are not classifiable into the universality class described by level dynamics. Nevertheless, the curvature distribution at tails shows an amazing universality. On the other hand, the kicked rotator is reduced to a generalized Sutherland system, and we can easily confirm universal behavior of $P(K)$. Takami and Hasegawa (1992) also analyzed a kicked rotator without time-reversal symmetry, verifying the validity of $P_{UE}(K)$ in (5.114). Thus the result in (5.114) works well even beyond the context of level dynamics.

Curvature distributions in small-K regions are given in fig. 5.10 in ordinary scales. A marked difference can be found near $K = 0$ between two systems. Unlike the mild hump at $K = 0$ for a kicked rotator, $P(K)$ for stadium billiards has a pronounced peak at $K = 0$, reflecting the existence of solitonic structures in fig. 5.8(*a*): an assembly of straight lines in fig. 5.8(*a*) increases the weight of distribution at $K = 0$. The behavior of $P(K)$ at $K = 0$ will thus be a convenient criterion by which to distinguish individual features of quantum analogs of chaotic systems.

In this way, statistical mechanics of a gCM system leads to universal

Fig. 5.9 Curvature distributions: (*a*) stadium billiard; (*b*) kicked rotator. The inset of each figure shows nearest-neighbor spacing distribution and Gaussian curves. (Courtesy of T. Takami and H. Hasegawa.)

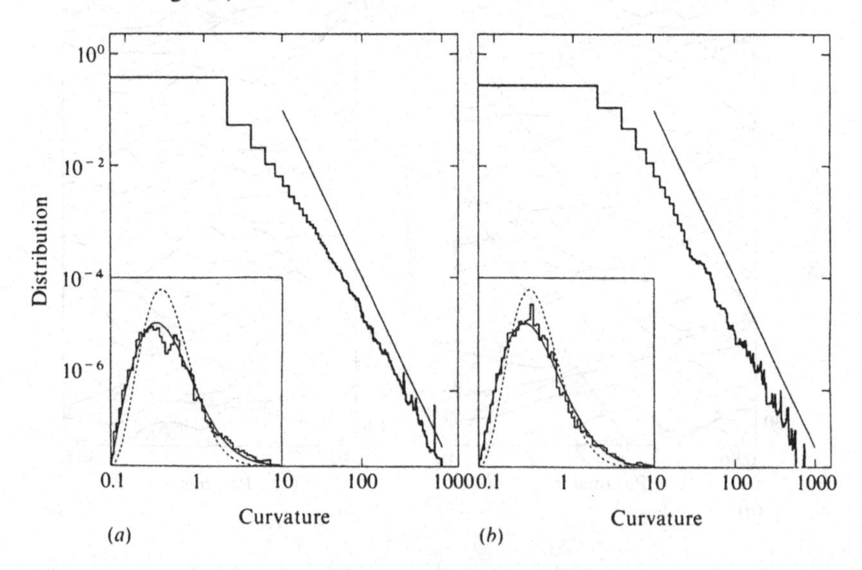

curvature distribution as well as the major results of random matrix theory, thereby suggesting a novel way to go beyond the traditional framework of random matrix theory.

5.6 Soliton-gas picture for quantum irregular spectra

Coherent structures are recognized as essential actors in natural science. One can mention the intricate role of dislocations in disordered materials (e.g., metallic glasses and amorphous silicon) and that of vortices in fluid turbulence. In closing this chapter we enrich this viewpoint by pointing out the role of solitons in quantum chaos.

As has been pointed out previously, generic features of quantum spectra for classically chaotic systems are much more fruitful than the results of random matrix theory. Even a coherent structure can coexist with irregular spectra for Gaussian orthogonal or unitary ensembles (GOE or GUE). The existence of solitonic structures (i.e., mobile avoided crossings) has been found by experiment on diamagnetic Rydberg atoms in fig. 5.4 as well as in numerical data for a pulsed spin system in fig. 5.2 and a stadium billiard in fig. 5.8(*a*). These solitons can be attributed to regular orbits in the underlying classical dynamics such as stable KAM tori (in generic systems) or unstable periodic orbits (in K systems). We now show theoretically how solitons can coexist with irregular level structures.

Fig. 5.10 Curvature distribution at small K values. Thin and thick lines correspond to stadium billiard and kicked rotator, respectively. (Courtesy of T. Takami and H. Hasegawa.)

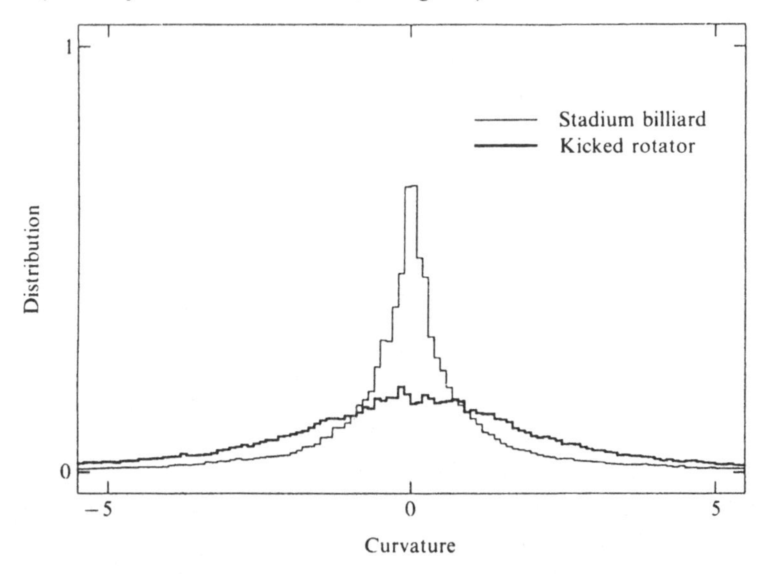

Let us consider the generic Hamiltonian (5.1). The study in section 5.2 revealed that the eigenvalue problem for (5.1) is reduced to the problem of solving the generalized Calogero–Moser (gCM) nonlinear dynamical system given by canonical equations (5.13) with the effective classical Hamiltonian (5.11). The most astonishing aspect of a gCM system is its complete integrability. This fact implies that, for any kind of initial values for $\{x_n\}$, we should inevitably see a soliton gas in large-τ regimes and, conversely, a soliton gas at $\tau = 0$ will yield complicated level structures in some nonvanishing τ regions. Solitons would therefore be compatible with GOE spectra.

To examine concretely our ideas, we employ a generalized Demkov–Osherov Hamiltonian (Demkov and Osherov, 1967) originally used for the problem of multidimensional Landau–Zener transitions:

$$H = \begin{bmatrix} A & C \\ C & B \end{bmatrix} \tag{5.115a}$$

with

$$A = \begin{bmatrix} p_1\tau + q_1 & w & w \\ w & p_2\tau + q_2 & w \\ w & w & \dots \end{bmatrix},$$

$$B = \begin{bmatrix} a_1 & 0 & 0 \\ 0 & a_2 & 0 \\ 0 & 0 & \dots \end{bmatrix}, \qquad C = \begin{bmatrix} u & u & . \\ u & u & . \\ . & . & . \end{bmatrix}, \tag{5.115b}$$

where $\{p_i\tau + q_i\}$ and $\{a_i\}$ denote solitonic (crossing) and horizontal (τ-independent) levels, respectively, and w and u are soliton–soliton and soliton–horizontal couplings, respectively. The model (5.115), which is a straightforward extension of (5.32) and (5.39), belongs to the category (5.1) and is especially advantageous for seeing solitons.

Single- and double-soliton profiles were shown in section 5.4. Collisions among four solitons are shown in fig. 5.11. It is found that solitons, when merged, become a seed of irregular energy spectra. By this observation, one is tempted to increase the number of solitons so that irregular spectra characteristic of quantum chaos are available. For given values u and w, $\{a_i\}$ and $\{q_i\}$ are chosen such that a Poisson level-spacing distribution is available at $\tau = 0$. Momenta of solitons, $\{p_i\}$, are assumed to obey the Maxwell–Boltzmann distribution $P(p)\,\mathrm{d}p \propto \exp(-p^2/2\kappa T)\,\mathrm{d}p$. Temperature T is therefore responsible for explicit forms of the perturbation V in (5.1). At $\tau = 0$, we prepare a set of 100 horizontal levels interspersed with a pair of sets each consisting of 130 solitonic levels. This arrangement is most suitable for the purpose of seeing collisions of solitons frequently. By diagonalizing the matrix in (5.115) at $\tau = 0$, we obtain a complete set of initial values for $\{x_n\}$ and $\{L|n\rangle\}$ and their canonical-conjugate variables. We proceed to solve gCM equation (5.13) for $\tau > 0$. Figure 5.12

Fig. 5.11 Collision among four solitons.

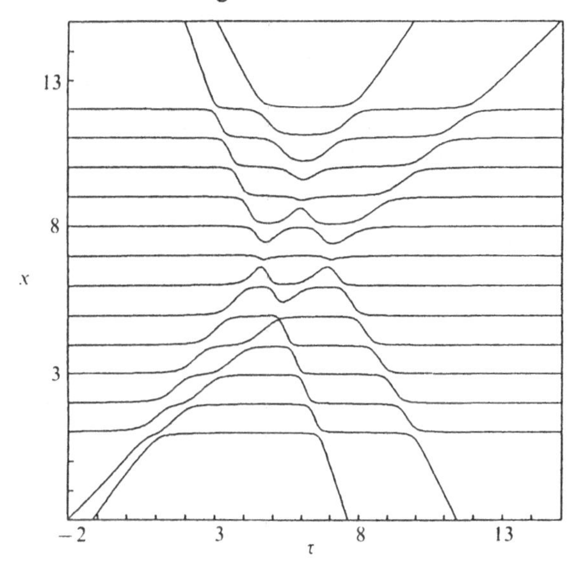

Fig. 5.12 Irregular spectrum caused by many solitons.

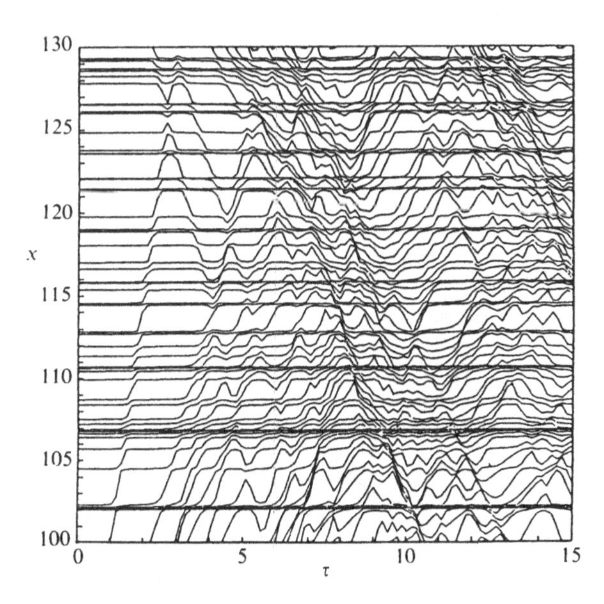

shows the τ-dependence of the energy spectrum in a fixed energy range where an assembly of 100 horizontal levels are located at $\tau = 0$. We see the following. (1) As τ increases, the invasion of many solitons into this region becomes noticeable and the number of collisions between solitons and horizontal levels is increased. Soliton structures in this region are similar to those in fig. 5.4. (2) For $\tau > \tau_c$ ($\tau_c \simeq 5.5$), solitons are well concentrated and an irregular feature prevails in the spectrum. (3) Nevertheless, even for $\tau > \tau_c$, some solitons retain their coherent profiles, showing coexistence of solitons with the irregular spectrum.

Corresponding level-spacing distributions at $\tau = 0$ and $\tau = 15$ are given in figs. 5.13(*a*) and 5.13(*b*), respectively. Figure 5.13(*a*) is the Poisson distribution characteristic of integrable systems. Figure 5.13(*b*) is

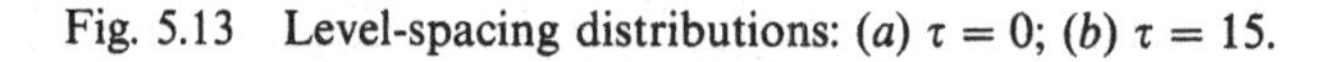

Fig. 5.13 Level-spacing distributions: (*a*) $\tau = 0$; (*b*) $\tau = 15$.

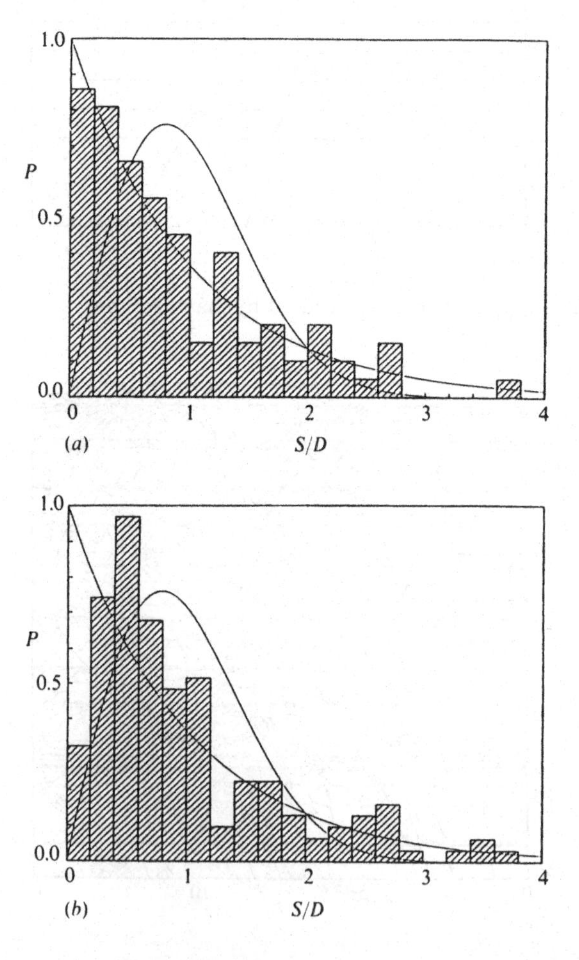

the Wigner distribution for GOE. For $0 < \tau < \tau_c$, we see intermediate distributions. The Wigner distribution is found to be observed for $\tau_c < \tau < \tau_c'$ ($\tau_c' \gg 15.0$ in the present case). Here τ_c' is the time when most solitons go away from the energy range in fig. 5.12, leaving the horizontal levels. Our study for other values of κT (i.e., global temperature) indicates that increase in κT widens the range in which solitons are well concentrated with a resultant decrease in τ_c and increase in $\tau_c' - \tau_c$. In other words, at low temperatures only a limited number of solitons can merge in any energy (ε) and parameter (τ) regions, thus failing to yield GOE. An extensive study using enlarged (2900×2900) matrices with 500 horizontal and 1200×2 solitonic levels (Ishio and Nakamura, 1992) has confirmed the universality of the above assertion.

In conclusion, a soliton-gas model gives a universal picture for energy spectra of nonintegrable and chaotic systems. Condensation of solitons leads to irregular spectra for GOE when the concentration of solitons is above its threshold. Thus, like vortices in fluid turbulence, solitons can play a crucial role in the field of quantum chaos and the soliton-gas picture provides a key to understanding both universality and nonuniversality aspects of quantum chaos. We have now liberated ourselves from the stubborn legacy of random matrix theories, by developing a new paradigm of nonlinear dynamics born of adiabatic-ansatz eigenvalue problems.

6
Nonadiabatic generalization, field-theoretical model and future prospects

In this final chapter we provide deep insights into other important problems of quantum chaos. So long as only one parameter is varied in Hamiltonians, desymmetrized energy spectra exhibit no degeneracies except for accidental ones and avoided crossings can be widely seen. When several parameters are changed, we see degeneracies in general. In this context, we choose two subjects which are fundamental for both quantum chaos and quantum mechanics itself. First, we explore the role of gauge structures in nonadiabatic transitions at avoided crossings. Local features of the hypothetical gauge potential are found to modify the nature of intersection of adiabatic energy surfaces and thereby affect the transition probability. Then, again imposing the adiabatic ansatz, eigenvalue problems for Hamiltonians with several nonintegrability parameters are analyzed, deriving a new field-theoretical model for interacting energy surfaces. The final section is devoted to my speculations on genuine quantum chaos and prospects of improving the ordinary framework of quantum mechanics.

6.1 Nonadiabatic transitions at avoided level crossings

The discovery of a quantum adiabatic phase (Berry, 1984) accompanying transport for closed paths has had a great impact on various fields of physics and chemistry. This global phase was originally discussed in connection with intersection of molecular energy surfaces. Its origin is nowadays attributed to connection in the Hilbert bundle, i.e., to the hypothetical gauge potential (see chapter 1).

A generalization of this kind of study to nonadiabatic transitions is much less trivial, because there is a possibility of seeing a local (rather than global) feature of the gauge potential via transition probability.

While several works have made a start along these lines, most of their efforts have been directed towards finding corrections to the global phase factor for closed or near-closed paths. Nonadiabatic transition occurs, however, quite widely for both open and closed paths and constitutes a key mechanism for understanding a variety of state-changing phenomena, such as atomic and molecular collisions, surface reactions and the solar neutrino puzzle. This transition, being induced effectively at avoided crossings (ACs), evidently shares a common stage with the adiabatic phase.

On the other hand, no book on quantum chaos would be complete without a description of nonadiabatic transitions at ACs. As noted in chapter 1, the nonintegrability and chaos in classical dynamics lead to absence of quantum numbers, thereby resulting in many ACs and associated irregular spectra. In other words, ACs signal an outbreak of quantum chaos. This is the reason why our effort so far has been devoted to studying statistical aspects of ACs such as level-spacing and curvature distributions. For the purpose of understanding ACs more deeply, however, singular structures around even a single AC should be explored in the case of nonadiabatic quantum transport processes. In fact, just as classical chaos appears in its temporal behavior, quantum chaos should expose its ingredients in its time dependence. The approach in this section is complementary to another kind of time-dependent treatment made on morphological transition in chapter 3.

We concentrate on the dynamics of 2×2 systems in a molecular reaction along open or closed paths; typically, when both translational and rotational degrees of freedom are coupled. While an electronic transition in molecular collisions will be chosen as a prototype, the theoretical results will be applicable to diverse other systems. Nuclear motion is treated classically. For a given nuclear path $\mathbf{R}(t)$ $(=(X(t), Y(t), Z(t)) = R(t)(\sin\Theta(t)\cos\Phi(t), \sin\Theta(t)\sin\Phi(t), \cos\Theta(t)))$, we evaluate the transition amplitude $K_{21}[\mathbf{R}(t)]$ from an electronic state 1 at initial time $t_1 = -\infty$ to another state 2 at final time $t_2 = +\infty$.

Let us choose a generic form for the $\mathbf{R}$-dependent electronic Hamiltonian,

$$H(\mathbf{R}(t)) = \frac{1}{2}\begin{bmatrix} Z(t) & X(t) - \mathrm{i}Y(t) \\ X(t) + \mathrm{i}Y(t) & -Z(t) \end{bmatrix}, \tag{6.1}$$

and define its adiabatic eigenstates and eigenvalues

$$H(\mathbf{R})\Psi_i(\mathbf{R}) = E_i(\mathbf{R})\Psi_i(\mathbf{R}) \qquad (i = 1, 2), \tag{6.2a}$$

where

$$E_1(\mathbf{R}) = -E_2(\mathbf{R}) = -R/2. \tag{6.2b}$$

Ψ_i are now complex functions. The time-dependent electronic wavefunction is given by the expansion $\Psi(t) = c_1(t)\Psi_1(\mathbf{R}) + c_2(t)\Psi_2(\mathbf{R})$. The Schrödinger equation $i\hbar\dot{\Psi}(t) = H(\mathbf{R})\Psi(t)$ can then be reduced to

$$i\hbar\dot{c}_1(t) = (E_1(t) - A_1(t))c_1(t) - i\hbar\langle\Psi_1|\dot{\Psi}_2\rangle c_2(t),$$
$$i\hbar\dot{c}_2(t) = (E_2(t) - A_2(t))c_2(t) - i\hbar\langle\Psi_2|\dot{\Psi}_1\rangle c_1(t), \tag{6.3a}$$

where

$$A_i(t) \equiv i\hbar\langle\Psi_i|\dot{\Psi}_i\rangle = \begin{cases}(\hbar/2)\dot{\Phi}(1 - \cos\Theta) & \text{for } i = 1\\ (\hbar/2)\dot{\Phi}(1 + \cos\Theta) & \text{for } i = 2.\end{cases} \tag{6.3b}$$

(Overdots imply time-derivatives.) $\langle\Psi_1|\dot{\Psi}_2\rangle = \langle\Psi_2|\dot{\Psi}_1\rangle = 0$ at $t = \pm\infty$ are assumed so as to ensure no coupling between $|\Psi_1\rangle$ and $|\Psi_2\rangle$ at $t = \pm\infty$. A_i in (6.3b) are hypothetical gauge potentials, which can lead to the novel global phase (Berry, 1984) for a closed path in $\mathbf{R}$ space in the adiabatic limit ($\langle\Psi_1|\dot{\Psi}_2\rangle = \langle\Psi_2|\dot{\Psi}_1\rangle = 0$). In general cases ($\langle\Psi_1|\dot{\Psi}_2\rangle \neq 0$, $\langle\Psi_2|\dot{\Psi}_1\rangle \neq 0$), we can see below a much more important role of gauge potential in nonadiabatic transition, irrespective of open or closed paths.

We solve (6.3a) with initial condition $c_1(t_1) = 1$, $c_2(t_1) = 0$. Let us take new coefficients $a_i(t) = c_i(t)\exp\{(i/\hbar)\int_{t_1}^{t} dt'\,(E_i(t') - A_i(t'))\}$ and introduce

$$\Delta E \equiv E_2(t) - E_1(t), \qquad \Delta A \equiv A_2(t) - A_1(t). \tag{6.4}$$

Equations (6.3a) then become valid for $a_1(t)$ and $a_2(t)$. We first apply a perturbational treatment which has the advantage of clearly elucidating the gauge structure. The first-order perturbation solution for $a_1(t)$ at $t = t_2$ is given by

$$a_2(t_2) = -\int_{t_1}^{t_2} dt\,\langle\Psi_2|\dot{\Psi}_1\rangle \exp\{(i/\hbar)\int_{t_1}^{t} dt'\,(\Delta E(t') - \Delta A(t'))\}.$$

Therefore

$$K_{21}[\mathbf{R}(t)] = c_2(t_2)$$
$$= -\int_{t_1}^{t_2} dt\,\langle\Psi_2|\dot{\Psi}_1\rangle \exp\Big\{(-i/\hbar)\int_{t_1}^{t} dt'\,(E_1(t') - A_1(t'))$$
$$- (i/\hbar)\int_{t}^{t_2} dt'\,(E_2(t') - A_2(t'))\Big\}. \tag{6.5}$$

Equation (6.5) indicates that transition is induced at time t with net transition amplitude given as a sum over all possible transition times. The integral in (6.5) can be evaluated by stationary phase methods. The

t-derivative of the phase $\{\cdots\}$ in (6.5) yields

$$\Delta E(t) - \Delta A(t) = 0. \tag{6.6}$$

For a slow change of $\Phi(t)$, (6.6) can be satisfied only for complex times because of the avoided crossings. Because of the absence of real-time solutions of (6.6), we look for a real time t_0 at which $\{\cdots\}$ in (6.5) is least-rapidly varying: t_0 is a solution of $(d/dt)[\Delta E(t) - \Delta A(t)] = 0$. Taking a Taylor expansion of the phase around t_0, (6.5) becomes

$$K_{21}[\mathbf{R}(t)] = -\langle\Psi_2|\dot{\Psi}_1\rangle_{t_0} \exp(-i\mathscr{I}/\hbar) \int_{-\infty}^{\infty} dt$$
$$\times \exp[(i/\hbar)(\Delta E_0 - \Delta A_0)(t - t_0) + (i/6\hbar)(\Delta\ddot{E}_0 - \Delta\ddot{A}_0)(t - t_0)^3] \tag{6.7}$$

with

$$\mathscr{I} = \int_{t_1}^{t_0} dt\,[E_1(t) - A_1(t)] + \int_{t_0}^{t_2} dt\,[E_2(t) - A_2(t)], \tag{6.8}$$

where ΔE_0, $\Delta\ddot{E}_0$, ΔA_0 and $\Delta\ddot{A}_0$ are values at $t = t_0$. The integral in (6.7) becomes the Airy function. Noting its asymptotic form and setting the pre-exponential factor to 1 (this point will be clarified later), we obtain the transition probability for a single passage through the avoided crossing:

$$P_{2\leftarrow 1} = |K_{21}[\mathbf{R}(t)]|^2$$
$$= \exp\{(-4(2)^{1/2}/3\hbar)(\Delta E_0 - \Delta A_0)[(\Delta E_0 - \Delta A_0)/(\Delta\ddot{E}_0 - \Delta\ddot{A}_0)]^{1/2}\}. \tag{6.9}$$

Equation (6.9) can be rewritten in a universal form. Let us expand (6.6) around t_0 as $\Delta E(t) - \Delta A(t) = \Delta E_0 - A_0 + (1/2)(\Delta\ddot{E}_0 - \Delta\ddot{A}_0)(t - t_0)^2$. For this expression, (6.6) has complex-time solutions

$$\tau_c = t_0 + i[2(\Delta E_0 - \Delta A_0)/(\Delta\ddot{E}_0 - \Delta\ddot{A}_0)]^{1/2} \tag{6.10}$$

and its complex conjugate τ_c^*. Then the action integral

$$(i/\hbar)\int_{\tau_c^*}^{\tau_c} dt\,(\Delta E - \Delta A) \tag{6.11}$$

is found to reproduce exactly the argument of the exponential in (6.9). This proof also indicates that the gauge potential induces shifts and bifurcations of branching points of adiabatic eigenvalues in the complex-time plane. The transition probability in (6.9) is thus written more compactly as

$$P_{2\leftarrow 1} = \exp\left\{(\mathrm{i}/\hbar)\int_{\tau_c^*}^{\tau_c} \mathrm{d}t\,[\Delta E(t) - \Delta A(t)]\right\}. \qquad (6.12)$$

In the absence of gauge potential ($\Delta A(t) = 0$), a well-known fact is that perturbation series solution culminates in (6.12) and that (6.12) is exact.

In our case of $\Delta A(t) \neq 0$, (6.12) is found to be exact for slow phase dynamics, i.e., when the characteristic time for phase motion $\Phi(t)$ is larger than other time scales. The proof is simple. Let us go back to the Schrödinger equation in a diabatic representation with original Hamiltonian (6.1),

$$\mathrm{i}\hbar\dot{\Psi} = H(\mathbf{R}(t))\Psi. \qquad (6.13)$$

Applying unitary transformation $\Psi = U\hat{\Psi}$ with

$$U = \begin{bmatrix} \mathrm{e}^{-\mathrm{i}\Phi/2} & 0 \\ 0 & \mathrm{e}^{\mathrm{i}\Phi/2} \end{bmatrix}, \qquad (6.14)$$

(6.13) is reduced to

$$\mathrm{i}\hbar\dot{\hat{\Psi}} = \tfrac{1}{2}\begin{bmatrix} R\cos\Theta - \hbar\dot{\Phi} & R\sin\Theta \\ R\sin\Theta & -(R\cos\Theta - \hbar\dot{\Phi}) \end{bmatrix}\hat{\Psi}. \qquad (6.15)$$

Since we shall move from the real-t axis to the complex-t plane, the Hamiltonian in (6.15) is assumed analytic throughout some strip of the complex-t plane centered on the real axis. We can then borrow the formula by Stueckelberg (1932), arriving at

$$P_{2\leftarrow 1} = \exp\left\{(\mathrm{i}/\hbar)\int_{\hat{t}_c^*}^{\hat{t}_c} \Delta\hat{E}\,\mathrm{d}t\right\}. \qquad (6.16a)$$

where

$$\Delta\hat{E} = \{(R\cos\Theta - \hbar\dot{\Phi})^2 + R^2\sin^2\Theta\}^{1/2}$$
$$= \{(\Delta E)^2 - 2\Delta E\cdot\Delta A + (A_1 + A_2)^2\}^{1/2} \qquad (6.16b)$$

and $\hat{t}_c$ is the branching point of $\Delta\hat{E}$ in (6.16b). For a slow phase dynamics ($\hbar\dot{\Phi} \ll \cos\Theta$), $\Delta\hat{E}$ and $\hat{t}_c$ are approximated as

$$\Delta\hat{E} \simeq \Delta E - \Delta A, \qquad (6.17a)$$

$$\hat{t}_c \simeq \tau_c, \qquad (6.17b)$$

and (6.16a) reduces to (6.12). So long as slow phase dynamics is guaranteed, (6.12) is valid even in fully nonadiabatic regions and τ_c is the branching point of $\Delta E - \Delta A$ rather than ΔE.

The crucial point is that formula (6.16) more perfectly incorporates the gauge potential. For illustration we give an example of the electronic transition when the underlying nucleus executes a transverse winding motion perpendicular to the Landau–Zener linear orbit:

$$R\cos\Theta = vt, \quad (6.18a)$$

$$R\sin\Theta = \Gamma, \quad (6.18b)$$

$$\Phi = (w/n)t^n \quad (6.18c)$$

with $n = 1, 2, 3, \ldots$. We have

$$\Delta\hat{E} = [\Gamma^2 + (vt - \hbar w t^{n-1})^2]^{1/2}. \quad (6.19)$$

For $\hbar w/v \ll 1$, we find that the original branching point $(\mathrm{i}\Gamma/v)$ moves to $\hat{t}_c = \mathrm{i}\Gamma/(v - \hbar w)$ for $n = 2$ and shows bifurcations for $n = 3, 4, \ldots$. (The $n = 1$ case is trivial because a suitable shift of time origin suppresses the effect of gauge potential in the quantum dynamics.) A schematic diagram for branching points is given in fig. 6.1.

While Berry (1990) considered the $n = 2$ case, his main concern was limited to a geometric amplitude factor near the adiabatic limit. So he underestimated the role of shifts and bifurcations of branching points caused by the gauge potential. By using formula (6.16), we can calculate analytically the transition rate so long as $\hbar w/v \ll 1$. (Strictly speaking, a more careful treatment is required for $n =$ odd cases when the effect of turning points is also essential (Child, 1974).) Figure 6.2 shows transition rates obtained by exact numerical iteration of (6.1) with (6.18) for various value n and w/v. Transition rates at $w/v = 0$ denote Landau–Zener universal ones (Landau, 1932; Zener, 1932):

$$P_{\mathrm{LZ}} = \exp[-\pi\Gamma^2/(2\hbar v)]. \quad (6.20)$$

Fig. 6.1 Branching points in complex-time plane: (*a*) shifts ($n = 2$); (*b*) bifurcations ($n = 3$).

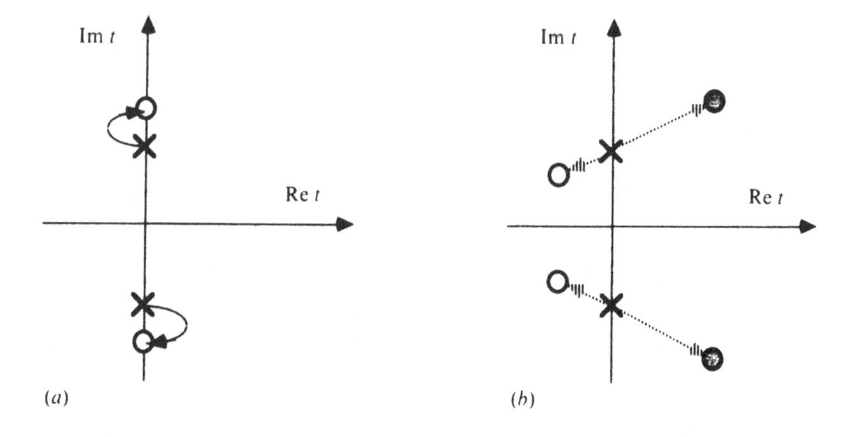

We recognize in fig. 6.2 that, as n increases, effects of shifts and bifurcations become crucial in applying formula (6.16).

The formulas in (6.16) are a generalization of the adiabatic theorem to contours in the complex-time plane. Paradoxically, the adiabatic theorem in the complex plane gives directly a nonadiabatic amplitude along the real axis. Also, we have revealed that the gauge potential, which led to the novel Berry phase for closed paths, can modify the nature of intersection of adiabatic energy surfaces and thereby affect crucially the nonadiabatic transition probability. All these findings are profoundly related to the genesis of the time-dependent Schrödinger equation where the first-order time derivative assumes the complex nature of wave-functions.

In general, (6.16) shows well not only for electronic transitions in molecular collisions but also for much more diverse systems in time-dependent external fields, when $\mathbf{R}(t)$ follows a curvilinear orbit in the neighborhood of avoided crossings that are the key element of quantum chaos.

6.2 Reduction to a field-theoretical model

Leaving the subject of nonadiabatic transitions at a single avoided crossing, we return to the level dynamics inspired by the adiabatic-ansatz

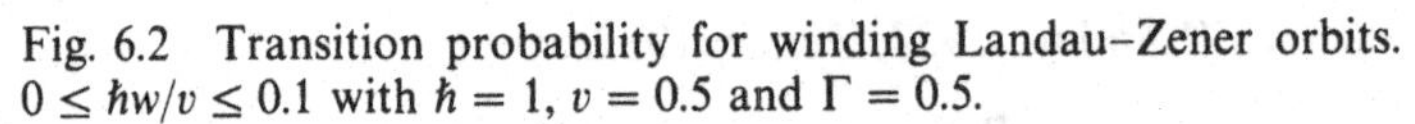

Fig. 6.2 Transition probability for winding Landau–Zener orbits. $0 \le \hbar w/v \le 0.1$ with $\hbar = 1$, $v = 0.5$ and $\Gamma = 0.5$.

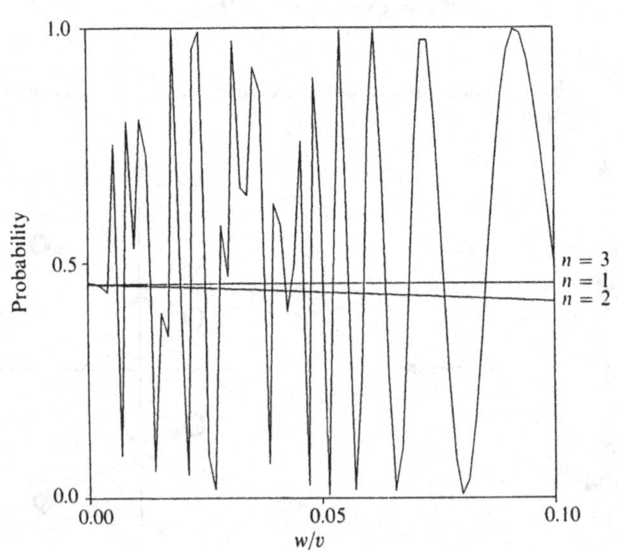

eigenvalue problem. Suppose several parameters are to be changed in the Hamiltonian, giving rise to degeneracies of eigenvalues in general. Crossings of eigenvalues have codimensions 2, 3 and 5 depending on the real, complex and quaternionic Hermitian nature of Hamiltonians, respectively (Arnold, 1978). The codimension here implies the number of independent parameters. In this case the fundamental equations of level dynamics in chapter 5 will be thoroughly altered (Nakamura, 1990; Nakamura *et al.*, 1992).

Let us choose the Hamiltonian in a form linear in nonintegrability parameters $\mathbf{x} = (x_1, x_2, \dots, x_d)$,

$$H(\mathbf{x}) = H_0 + \sum_{\mu=1}^{d} x_\mu V_\mu. \tag{6.21}$$

On the right-hand side of (6.21), the first term represents the integrable part and the second consists of a set of $d(\geq 2)$ nonintegrable perturbations describing, for instance, anisotropy energy, couplings with magnetic and electric fields, etc. Furthermore, we assume the noncommutativity relations

$$[H_0, V_\mu] = G_\mu \tag{6.22a}$$

$$[V_\mu, V_\nu] = \mathcal{G}_{\mu\nu} \tag{6.22b}$$

and Hermitian properties

$$H_0^\dagger = H_0, \tag{6.23a}$$

$$V_\mu^\dagger = V_\mu. \tag{6.23b}$$

From (6.22) and (6.23), G_μ and $\mathcal{G}_{\mu\nu}$ prove to be anti-Hermitian operators.

Consider the desymmetrized spectra corresponding to the eigenvalue problem

$$H(\mathbf{x})|n(\mathbf{x})\rangle = \phi_n(\mathbf{x})|n(\mathbf{x})\rangle. \tag{6.24}$$

The spectrum is now constructed in $\mathbf{x}$ space and each eigenvalue $\{\phi_n(\mathbf{x})\}$ defines a hyper-energy surface as in fig. 1.4. We can see contacts of adjacent energy surfaces at degenerate points besides avoided crossings of these surfaces. In the vicinity of each degeneracy, a pair of energy surfaces constitute two sheets of a double cone (diabolo). (See fig. 1.3, where $\{E_n\}$ and $\{R_\mu\}$ should be read as $\{\phi_n\}$ and $\{x_\mu\}$, respectively.) The diabolical point is a source of topological singularity: for a closed loop around the degeneracy in $\mathbf{x}$ space, a pair of eigenstates $|n(\mathbf{x})\rangle$ and $|n+1(\mathbf{x})\rangle$ joining the double cone acquire the global Berry phase (Shapere and Wilczek, 1989)

$$\mathrm{i} \sum_{\mu=1}^{d} \oint \langle n(\mathbf{x})|\partial_\mu|n(\mathbf{x})\rangle \, \mathrm{d}x_\mu. \tag{6.25}$$

Formula (6.25) also applies to the $|n+1(\mathbf{x})\rangle$ state. The global phase (6.25) is due to the phase arbitrariness of eigenfunctions, i.e., an essential characteristic of the traditional framework of quantum mechanics. This manifests the fact that eigenfunctions are multivalued in a higher-dimensional parameter space, though of course single-valued in the configuration space. The integrand in (6.25), $\mathrm{i}\langle n(\mathbf{x})|\partial_\mu|n(\mathbf{x})\rangle$, plays the role of gauge potential as is shown below. As a natural extension of thought on the single-parameter case, one may anticipate that the presence of many diabolos and avoided crossings of hyper-energy surfaces would be an indicator of quantum chaos characterized by several nonintegrability parameters. Taking $\{x_n\}$ as d-dimensional Euclidean 'space-time' coordinates, we regard $\{\phi_n(\mathbf{x})\}$ and $\{|n(\mathbf{x})\rangle\}$ $(=(|1(\mathbf{x})\rangle, |2(\mathbf{x})\rangle, \ldots, |D(\mathbf{x})\rangle)$ with $\langle n(\mathbf{x})|n'(\mathbf{x})\rangle = \delta_{nn'})$ as *internal* classical vector and matrix fields, respectively. Note that $1 \le \mu \le d$; $1 \le n \le D$ and D may be infinite. Einstein's contraction is *not* used. By taking the $\mathbf{x}$ derivative of (6.24) we have, following the method in section 5.2,

$$\partial_\mu^2 \phi_n = 2 \sum_{m(\neq n)} \langle n|V_\mu|m\rangle\langle m|V_\mu|n\rangle(\phi_n - \phi_m)^{-1}, \tag{6.26a}$$

$$\partial_\mu|n\rangle = \sum_{m(\neq n)} |m\rangle\langle m|V_\mu|n\rangle(\phi_n - \phi_m)^{-1} - \mathrm{i}A_{n,\mu}|n\rangle, \tag{6.26b}$$

where

$$A_{n,\mu} = \mathrm{i}\langle n|\partial_\mu|n\rangle. \tag{6.27}$$

On introducing a set of constant orthonormal vectors $\mathbf{e}_1, \mathbf{e}_2, \ldots, \mathbf{e}_d$ with $\mathbf{e}_\mu \cdot \mathbf{e}_\nu = \delta_{\mu\nu}$, (6.26) can be readily summarized in a convenient form as

$$\sum_\mu \partial_\mu^2 \phi_n \mathbf{e}_\mu = 2 \sum_\mu \sum_{m(\neq n)} \langle n|\bar{\Lambda}_\mu|m\rangle\langle m|\bar{\Lambda}_\mu|n\rangle \times (\phi_n - \phi_m)^{-3}\mathbf{e}_\mu, \tag{6.28a}$$

$$\sum_\mu (\partial_\mu + \mathrm{i}A_{n,\mu})\bar{\Lambda}_\mu|n\rangle\mathbf{e}_\mu = -\mathrm{i}\sum_\mu \sum_{m(\neq n)} \bar{\Lambda}_\mu|m\rangle\langle m|\bar{\Lambda}_\mu|n\rangle(\phi_n - \phi_m)^{-2}\mathbf{e}_\mu, \tag{6.28b}$$

with the non-negative Hermitian $\bar{\Lambda}_\mu = \Lambda_\mu - \lambda_\mu I$ which has its unique inverse $\bar{\Lambda}_\mu^{-1}$. Here

$$\begin{aligned} \Lambda_\mu &\equiv \mathrm{i}^{-1}[H, V_\mu] \\ &= \mathrm{i}^{-1}G_\mu + \mathrm{i}^{-1}\sum_{\nu\neq\mu} x_\nu \mathscr{G}_{\nu\mu} \end{aligned} \tag{6.29}$$

and λ_μ is the lowest eigenvalue of Λ_μ (see section 5.2). Since Λ_μ in (6.29) does not include x_μ, both λ_μ and $\bar{\Lambda}_\mu$ are x_μ-independent so that $\partial_\mu \bar{\Lambda}_\mu = 0$, which has been exploited in deriving (6.28*b*). For the description below we should also note that $\langle n|\bar{\Lambda}_\mu|n\rangle = -\lambda_\mu$, independent of *n*. Furthermore, let us make a comment on $A_{n,\mu}$ defined in (6.27). Contrary to the single-parameter case, $A_{n,v}$ is nonvanishing in general and behaves like a gauge potential. Under the gauge transformation $|n\rangle \rightarrow |n\rangle \cdot \exp(\mathrm{i}\alpha_n)$, $A_{n,\mu}$ transforms as $A_{n,\mu} \rightarrow A_{n,\mu} - \hbar\partial_\mu\alpha_n$, but the 'magnetic field'

$$B_{n,\mu\nu} \equiv \partial_\mu A_{n,\nu} - \partial_\nu A_{n,\nu} \tag{6.30}$$

is gauge-invariant. The comparison with existing results is straightforward: for $d = 1$ with $A_{n,\mu} = 0$, (6.28) reduces to the gCM system, while, for $d > 1$ and suppressing the level dynamics (6.28), only the gauge potential (6.27) is available (Shapere and Wilczek, 1989).

Note that (6.28) with (6.27) can be obtained from a field-theoretical model. In fact, if we introduce a classical Lagrangean density

$$\mathscr{L} = \mathscr{L}_0 + \mathscr{L}', \tag{6.31a}$$

with

$$\mathscr{L}_0 = \sum_{\mu,n} \tfrac{1}{2}(\partial_\mu\phi_n)^2\mathbf{e}_\mu - \mathrm{i}\sum_{\mu,n}(\partial_\mu\langle n|)\bar{\Lambda}_\mu|n\rangle\mathbf{e}_\mu - \sum_{\mu,\nu} A_{n,\mu}(\langle n|\bar{\Lambda}_\mu|n\rangle + \lambda_\mu)\mathbf{e}_\mu \tag{6.31b}$$

and

$$\mathscr{L}' = -\frac{1}{2}\sum_\mu \sum_{\substack{m,n\\(m\neq n)}} \langle n|\bar{\Lambda}_\mu|m\rangle\langle m|\bar{\Lambda}_\mu|n\rangle \times (\phi_n - \phi_m)^{-2}\mathbf{e}_\mu, \tag{6.31c}$$

we derive the Euler–Lagrange equations from (6.31) as

$$\partial\mathscr{L}/\partial\phi_n - \sum_\mu \partial_\mu(\partial\mathscr{L}/\partial_\mu\phi_n) = 0, \tag{6.32a}$$

$$\partial\mathscr{L}/\partial\langle n| - \sum_\mu \partial_\mu(\partial\mathscr{L}/\partial_\mu\langle n|) = 0 \tag{6.32b}$$

and

$$\partial\mathscr{L}/\partial A_{n,\mu} = 0. \tag{6.33}$$

As can easily be verified, (6.32*a*) and (6.32*b*) exactly reproduce (6.28*a*) and (6.28*b*), respectively. On the other hand, (6.33) yields the constraint given below (6.29). Noting that $\partial_\mu(\bar{\Lambda}_\mu|n\rangle) = \bar{\Lambda}_\mu\partial_\mu|n\rangle$ and multiplying (6.32*b*) or (6.28*b*) from the left by $\mathbf{e}_\nu\langle n|\bar{\Lambda}_\nu^{-1}$, we also recover (6.27).

In contrast with the generalized Calogero–Moser model in 1 + 1 dimensions for the single-parameter case, we now have a novel field-theoretical model inspired by the eigenvalue problem associated with several nonintegrability parameters. In this model, 'space–time' coordinates represent a set of nonintegrability parameters, and matter fields ($\{\phi_n\}$ and $\{|n\rangle\}$) are coupled with apparent gauge fields $\{A_{n,\mu}\}$. The gauge fields here are composed of the matter fields themselves, as in the complex Grassmannian sigma model. Consequently, the 'magnetic field' in (6.30) is provided by

$$B_{n,\mu\nu} = \sum_{m(\neq n)} \{\langle n|\bar{\Lambda}_\mu|m\rangle\langle m|\bar{\Lambda}_\nu|n\rangle \times (\phi_n - \phi_m)^{-4} - \text{H.c.}\}. \quad (6.34)$$

We can proceed to search for instanton-like structures in (6.32) and (6.33) and to construct the statistical mechanics (6.31), just as attempted for the gCM system in the previous chapter.

The present scheme is also applicable to generalized kicked rotators with several nonintegrability parameters. For a quantum Hamiltonian

$$H = H_0 + \sum_{\mu=1}^{d} x_\mu \hat{V}_\mu \sum_{j=-\infty}^{\infty} \delta(t - 2\pi j), \quad (6.35)$$

quasi-eigenvalues $\{\phi_n\}$ and quasi-eigenstates $\{|n\rangle\}$ satisfy the eigenvalue problem for a one-period unitary operator (see section 5.3):

$$U|n\rangle \equiv \left\{\exp\left(-\mathrm{i}\sum_{\mu=1}^{d} x_\mu V_\mu\right)\right\} U_0|n\rangle = \exp(-\mathrm{i}\phi_n)\,|n\rangle, \quad (6.36)$$

where $V_\mu = \hat{V}_\mu/\hbar$ and $U_0 = \exp[(-\mathrm{i}/\hbar)2\pi H_0]$. Suppose that the case with (6.22*a*) and $[V_\mu, V_\nu] = c_{\mu\nu}$, where $\{c_{\mu\nu}\}$ are pure-imaginary *c* numbers, is chosen. Then the scenario from (6.26) through (6.34) holds in this case also, except for the following points: (i) $(\phi_n - \phi_m)$ is replaced here by $2\sin[(\phi_n - \phi_m)/2]$; (ii) the alternative definition for Λ_μ in (6.29) is

$$\begin{aligned}\Lambda_\mu &\equiv V_\mu - U^\dagger V_\mu U \\ &= V_\mu - U_0^\dagger\left\{\exp\left(\mathrm{i}\sum_{\nu\neq\mu} x_\nu V_\nu\right)\right\} V_\mu \left\{\exp\left(-\mathrm{i}\sum_{\kappa\neq\mu} x_\kappa V_\kappa\right)\right\} U_0,\end{aligned}$$

which is again x_μ-independent. (In deriving the last expression, we have used the equality

$$\begin{aligned}\exp\left\{-\mathrm{i}x_\mu V_\mu - \mathrm{i}\sum_{\nu\neq\mu} x_\nu V_\nu\right\} &= \exp(-\mathrm{i}x_\mu V_\mu)\exp\left\{-\mathrm{i}\sum_{\nu\neq\mu} x_\nu V_\nu\right\} \\ &\quad \times \exp\left\{2^{-1}\sum_{\nu\neq\mu} x_\mu x_\nu c_{\mu\nu}\right\}\end{aligned}$$

and its Hermitian conjugate.) Consequently, we obtain a field-theoretical counterpart of the generalized Sutherland system.

The theory developed here is beyond the scope of the theory of surfaces in d dimensions in Riemannian geometry. In the latter, fundamental equations such as the Gauss–Weingarten equations are associated with a single surface, while, in the theory developed here, mutually interacting surfaces are treated. The study of quantum chaos thus opens a new avenue for a field theory of interacting energy surfaces.

The fictitious gauge potential described so far is worth scrutinizing further. It is found to deform quantum-mechanical commutation relations for slow (nuclear) variables (Jackiw, 1988; Shapere and Wilczek, 1989). Some discussion of this interesting possibility is presented before moving to the closing section.

Suppose that a Hamiltonian for molecules with slow (nuclear) **P**, **R** and fast (electronic) **p**, **r** variables is

$$H = \mathbf{P}^2/2M + \mathbf{p}^2/2m + V(\mathbf{R}, \mathbf{r}), \tag{6.37}$$

and apply the Born–Oppenheimer approximation to the complete eigenvalue problem $H\Psi = E\Psi$. If one chooses an eigenfunction in a factorized form

$$\Psi(\mathbf{R}, \mathbf{r}) = \Phi(\mathbf{R})|n; \mathbf{R}\rangle, \tag{6.38}$$

one obtains for $\Phi(\mathbf{R})$, when $|n; \mathbf{R}\rangle$ is nondegenerate, the equation

$$\{(\mathbf{P} - \mathbf{A})^2/2M + V_{\text{eff}}(\mathbf{R})\}\Phi(\mathbf{R}) = E\Phi(\mathbf{R}) \tag{6.39a}$$

with

$$\mathbf{A}(\mathbf{R}) \equiv \mathrm{i}\langle n; \mathbf{R}|\nabla_{\mathbf{R}}|n; \mathbf{R}\rangle, \tag{6.39b}$$

$$V_{\text{eff}}(\mathbf{R}) = \varepsilon_n(\mathbf{R}) + (\nabla_{\mathbf{R}}\langle n; \mathbf{R}|\nabla_{\mathbf{R}}|n; \mathbf{R}\rangle - \mathbf{A}^2)/2M. \tag{6.39c}$$

The effective potential $V_{\text{eff}}(\mathbf{R})$ in (6.39*c*) consists of the adiabatic eigenvalue at **R** and a correction due to slow motion. The kinetic energy in (6.39*a*) resembles that of a particle in the presence of a gauge potential (see chapter 2). As a result we have the anomalous commutation relation for slow variables

$$[P_i, P_j] = \mathrm{i}\hbar\varepsilon^{ijk}B_k \tag{6.40}$$

with $\mathbf{B} = \nabla_{\mathbf{R}} \times \mathbf{A}$. In the anomaly phenomena in modern quantum field theory, some symmetries of classical physics may disappear because a symmetry-violating process occurs in quantum-mechanical treatment. The result (6.40) provides a typical example of this phenomenon.

In the field of quantum chaos, many diabolos as well as avoided crossings appear, which affect the fundamental commutation relations for underlying slow variables that have mostly been treated as external parameters in this book. In general, whenever one encounters the problem of separating fast (internal) from slow (external) variables, one may anticipate the presence of anomalous commutation relations for fast variables as a reaction to the anomalies in slow variables.

6.3 Future prospects

6.3.1 Continuing investigations of the trace formula

Before embarking upon conclusive assertions, we should return to the subject of semiclassical quantization in section 1.3 and comments on persistent challenges to improving the trace formula.

So long as one stays within the framework of the Schrödinger–Feynman quantum mechanics, Gutzwiller's trace formula remains one of the most valuable procedures for exposing the ambiguities obscuring the borderline between quantum and classical mechanics for chaotic systems. As emphasized in chapter 1, however, serious problems concerning Gutzwiller's trace formula emerge from its nonconvergence (for real energies) due to exponential proliferation of periodic orbits. This problem can be partly overcome either by smoothing the density of states or by analytic continuation of the trace formula. With the smoothing technique, Aurich *et al.* (1988) and Wintgen (1988) succeeded in obtaining the lowest few eigenstates for chaotic billiards and a hydrogen atom in a magnetic field, respectively. There remains, however, a more troublesome problem: the eigenvalues computed from the trace formula are not real! To solve this problem, new developments appeal to the theory of Riemann's zeta function together with invention of a novel resumation of series expansion called a Riemann–Siegel type resurgence (Berry and Keating, 1990). Only a brief discussion of this work is given here.

To begin with, define Riemann's zeta function

$$\zeta(z) \equiv \sum_{n=1}^{\infty} n^{-z} = \prod_{p} (1 - p^{-z})^{-1}, \tag{6.41}$$

where p runs over all prime numbers. The first and second definitions in (6.41) are given by Dirichlet series and Euler products, respectively, and $\zeta(z)$ is defined over the full complex-z plane by analytical continuation beyond the originally defined region (namely $\mathrm{Re}\, z > 1$). Riemann's famous hypothesis claims that zeros of $\zeta(z)$ are given by $z_n = \frac{1}{2} + \mathrm{i}E_n$ with all E_n real. Among many trials to prove this hypothesis, recent numerical computations by Odlyzko (1987) are most interesting. These computations showed that short-range level statistics of $\{E_n\}$ obey GUE, i.e., there is a universal level-spacing distribution for the system without time-reversal symmetry. Then it is tempting to construct a Gutzwiller-like trace formula whose poles yield $\{E_n\}$.

Noting that, on moving to the right of critical line $\mathrm{Re}\, z = \frac{1}{2}$, $\ln \zeta(z)$ increases by $i\pi$ at every zero E_n, the density of zeros $\rho(E)$ is

$$\rho(E) - \rho_0(E) = \frac{1}{\pi}\frac{\mathrm{d}}{\mathrm{d}E}\,\mathrm{Im}\ln\left[\zeta(\tfrac{1}{2} + \mathrm{i}E)\right]. \tag{6.42}$$

Using in (6.42) the Euler product for $\zeta(z)$, we obtain

$$\rho(E) - \rho_0(E) = -2\pi^{-1}\sum_p \sum_{k=-\infty}^{\infty}{}' (\ln p)\exp\left[-\tfrac{1}{2}|k|\ln p\right] \times \exp(\mathrm{i}Ek\ln p), \tag{6.43}$$

where $\rho_0(E) = (2\pi)^{-1}\ln(E/2\pi)$ is the mean density of zeros. In (6.43) $\sum'$ denotes absence of the $k = 0$ term in the summation. If one were to consider introducing a fictitious action

$$S(E) = Ek\ln p, \tag{6.44a}$$

then the fictitious period $T(E)$ and amplitude $A(E)$ would be given by

$$T(E) = \mathrm{d}S/\mathrm{d}E = k\ln p \tag{6.44b}$$

and

$$A(E) = \ln p\exp(-2^{-1}|k|\ln p) = T/k(\exp|k|\ln p)^{-1/2}. \tag{6.44c}$$

It is very interesting that, rewritten in terms of S, T and A, (6.43) takes the form of Gutzwiller's trace formula. For more details, see Berry (1986).

While $\{E_n\}$ available from $\zeta(z)$ have nothing to do with eigenvalues of any quantum-mechanical system, Selberg (1956) had previously established an elegant trace formula by invention of a new zeta function. Selberg's zeta function has zeros yielding eigenvalues of a quantum-mechanical particle sliding freely on a compact surface of constant negative curvature, with its classical counterpart fully chaotic.

Careful insight into the work of Riemann and Berry will provide a way to overcome the annoying problem of nonconvergence. The Riemann–Berry trace formulas indeed have the same weakness as Gutzwiller's formula: the periodic orbit sum is divergent due to exponential proliferation of periodic orbits dominating the amplitude factors. Nonetheless, Riemann's zeta function undoubtedly has distinctive real energies. Therefore, by manipulating the zeta function rather than the trace formula itself, one can obtain accurate real E values. In fact, Riemann's zeta function $\zeta(z)$, though defined in terms of divergent infinite Dirichlet series, can be transformed into a finite sum as

$$\zeta(\tfrac{1}{2} + \mathrm{i}E) \simeq 2 \exp\{-\pi \mathrm{i} N(E)\} \sum_{m=1}^{(E/2\pi)^{1/2}} m^{-1/2} \cos(E \ln m - \pi N(E)) \tag{6.45}$$

with $N(E) \cong (E/2\pi) \ln (E/2\pi e) + \frac{7}{8}$. (6.45) is called the Riemann–Siegel resurgence formula and, from its zeros, real E values can be computed with no serious problems.

The above interesting idea can be applied to Gutzwiller's trace formula. By simple integration and exponentiation of (1.34) together with an expansion $[2 \sinh (lu_\alpha/2)]^{-1} = \sum_{k=0}^{\infty} \exp[-l(k + \frac{1}{2})u_\alpha]$ in (1.35), one obtains (Gutzwiller, 1990)

$$\mathrm{Tr}\, G(E) - \mathrm{Tr}\, G_0(E) \simeq \frac{\mathrm{d}}{\mathrm{d}E} \ln Z(E), \tag{6.46}$$

where $Z(E)$ is a kind of zeta function defined as

$$Z(E) = \prod_{\alpha} \prod_{k=0}^{\infty} (1 - t_\alpha \Lambda_\alpha^{-k}) \tag{6.47a}$$

with quantum weights

$$t_\alpha = |\Lambda_\alpha|^{-1/2} \exp[\mathrm{i}(S_\alpha/\hbar - m_\alpha \pi/2)]. \tag{6.47b}$$

$\Lambda_\alpha (= \pm \exp(u_\alpha)$ or $\exp(\mathrm{i}u_\alpha)$ with $u_\alpha > 0$ for unstable and stable periodic orbits, respectively) are eigenvalues of a linearized Poincaré map around the orbit (see section 1.3). From (6.46), one recognizes that zeros of $Z(E)$ lead to the quantum eigenvalue. For generic Hamiltonian systems (6.46) is valid to leading order in $\hbar$, though it is exact for Selberg's geodesic flow on a surface of constant negative curvature. With a suitable expansion of (6.47) and grouping the resulting terms in ascending order of symbolic length, (6.47) may be resumed by the Riemann–Siegel look-alike resurgence (Berry and Keating, 1990). Here another resurgence called cycle expansions (Cvitanović and Eckhardt, 1989) would also be helpful.

Tanner *et al.* (1991) applied formulas (6.46) and (6.47) to the anisotropic Kepler problem and bounded billiards, obtaining fairly high-lying individual eigenvalues. For general Hamiltonian systems, however, complexity in symbolic coding of periodic orbits prevents us manipulating formulas (6.46) and (6.47). So it remains questionable whether zeta functions inspired by the trace formula are crucial in quantization of generic systems. Further, we should make some far more important remarks. Gutzwiller's trace formula is valid up to leading order in $\hbar$. Therefore, neither smoothing of level density nor resurgence for suitable zeta function can yield eigenvalues with precisions higher than the order of $\hbar^N$. These various revisions will be truly meaningful when combined with improvement of the trace formula so as to incorporate all order terms in $\hbar$. Such attempts, however, will lead us into a forest of complicated mathematics where we shall be lost, and the criticism raised in section 1.3 may be justified. One may envisage a possibility of constructing a nonlinear variant of quantum mechanics that could yield a successful one-to-one correspondence between quantum eigenvalues (eigenstates) and classical chaos.

6.3.2 Summary and future prospects

For the purpose of exploring quantum-mechanical manifestations of classical chaos, we have so far investigated both adiabatic-ansatz eigenvalue problems and quantum dynamics in the semiclassical region. Both bounded and open systems have been chosen from diverse branches of solid-state science. Well-known concepts such as diamagnetism, antiferromagnetism, spinwaves and electric conductance have been considered in terms of quantum chaos. The spectroscopy of highly-excited diamagnetic Rydberg atoms (Kleppner *et al.*, 1983) and the ionization of highly-excited hydrogen atoms by a microwave field (Bayfield and Koch, 1974; Leopold and Percival, 1978), which are not touched on in this book, are also being intensively studied in the context of quantum chaos.

In the framework of quantum mechanics, nonlinear aspects possessed by classical dynamical systems are fully incorporated into the Hamiltonian. Consequently eigenvalues and eigenfunctions are destined to bear random features. In fact, they have demonstrated erratic and irregular behaviors as encountered in random or disordered systems. The symmetry-induced quantum numbers involved in integrable systems become lost when nonintegrability and chaos are brought in. The resultant appearance of abundant avoided level crossings has proved to be the underlying mechanism for irregular energy spectra and wavefunction features. In particular, the sensitivity of eigenvalues to variation of nonintegrability parameters is quantified in terms of curvatures or susceptibility. The most important issue is the discovery of a completely integrable generalized Calogero–Moser system lying behind the adiabatic-ansatz eigenvalue problem. This system constitutes a new paradigm of nonlinear dynamics, since it embodies all the information on eigenvalues and eigenfunctions inherent in quantum chaos, and further, its statistical mechanics indicates a way of going beyond random matrix theory. A deep insight has also been given into gauge structure at diabolical conical intersections appearing when several nonintegrability parameters are operative. From the corresponding adiabatic-ansatz eigenvalue problem, we have even derived a field-theoretical model.

Despite the various significant findings described so far, however, the most critical question remains unanswered: how can chaotic behavior characterized by positive Lyapunov exponent, etc. show up in quantum dynamics? While interesting multifractal structures of spin wavefunction have been displayed (see chapter 3), suppression of chaotic diffusion is inevitable after the cross-over time. This point was also noted in the case of a kicked rotator (see chapter 1), where suppression of chaos was interpreted as being caused by a localization in momentum space. Eventually, time evolution of wavefunctions for bounded systems, whether driven or not, whether small or large in particle numbers, can indicate neither mixing nor Bernoulli properties. Nonetheless, our rational intelligence demands a genuine quantum chaos characterized by standard diagnostic criteria as used for classical chaos.

The absence of chaos (e.g., vanishing K entropy) even in the semiclassical limit of quantum dynamics is quite puzzling, since we know that in the limit $\hbar \to 0$ the time-dependent Schrödinger equation is converted into the Hamilton–Jacobi equation which may be chaotic in general. What clue is there to help solve the puzzle?

To take a step further towards the answer, one should sketch the philosophical background of the time-dependent Schrödinger equation. A Schrödinger equation can never be the equation describing de Broglie's matter wave, but rather has its logical foundation on Heisenberg's matrix equation for noncommutable canonical operators. This means that $|\Psi|^2$ does not describe field intensity at some positions. Instead, it represents a probability amplitude for an assembly of particles to have a given configuration. Let us consider a system of two interacting particles. Suppose that Φ describes a matter wave; it should then be a function of one space coordinates $\mathbf{q}$, satisfying

$$i\hbar \frac{\partial \Phi(\mathbf{q})}{\partial t} = \left\{-(\hbar^2/2m)\nabla_{\mathbf{q}}^2 + V(\mathbf{q}) + \int |\Phi(\mathbf{q}')|^2 U(\mathbf{q}', \mathbf{q})\, d^3\mathbf{q}'\right\}\Phi(\mathbf{q}). \quad (6.48)$$

The last term in $\{\cdots\}$ denotes the inter-particle interaction. Equation (6.48) is an integro-differential equation highly nonlinear in Φ, and therefore has a high probability of showing chaos. However, (6.48) cannot be a correct Schrödinger equation, but merely a classical field equation. (Although the field-quantization procedure will change (6.48) into a linear Schrödinger equation for interacting bosons, this is a different story.) Needless to say, the precise Schrödinger equation for the two-particle system is given by

$$i\hbar \frac{\partial}{\partial t} \Psi(\mathbf{q}_1, \mathbf{q}_2) = \left\{ \sum_{j=1}^{2} \left(-\frac{\hbar^2}{2m_j} \nabla_{\mathbf{q}j}^2 + V(\mathbf{q}_j) \right) + U(\mathbf{q}_1, \mathbf{q}_2) \right\} \Psi(\mathbf{q}_1, \mathbf{q}_2). \quad (6.49)$$

$|\Psi|^2$ provides the probability of finding particles 1 and 2 at $\mathbf{q}_1$ and $\mathbf{q}_2$, respectively. Let us assume (6.48) to be an alternative to the fundamental equation of quantum dynamics. We may then arrive at plausible chaos with nonvanishing K entropy, but neither a well-defined frequency nor an eigenvalue are available. The most serious disadvantage of the nonlinear form (6.48) is that one must sacrifice the great advantage of quantum mechanics, i.e., the superposition principle, for nonintegrability and chaos.

From the viewpoint of nonlinear dynamics, the content of quantum mechanics is stated as follows. All nonlinearity arising from coupling between particles and on-site potentials, etc. is absorbed into the Hamiltonian and therefore eigenfunction Ψ itself is forced to obey a linear equation like (6.49). Instead of N coupled nonlinear equations as in classical mechanics, we have only a linear equation for $\Psi(\mathbf{q}_1, \mathbf{q}_2, \ldots, \mathbf{q}_N)$ (e.g., in the case of $\mathbf{q}$ representation). Weinberg (1989) recently tried to make the Schrödinger equation nonlinear in his analysis of experimental results on precession frequency fluctuation of a $^9Be^+$ nucleus in a magnetic field. He assumed that the time dependence of the wavefunction (for a state l) is given by

$$i\hbar \frac{\mathrm{d}}{\mathrm{d}t} \Psi_l = \frac{\partial \mathscr{H}(\Psi, \Psi^*)}{\partial \Psi_l^*}, \quad (6.50)$$

where $\mathscr{H}$ is a real function of Ψ and Ψ^* satisfying the homogeneity requirement $\mathscr{H}(\lambda\Psi, \Psi^*) = \mathscr{H}(\Psi, \lambda\Psi^*) = \lambda\mathscr{H}(\Psi, \Psi^*)$ for any complex λ. Homogeneity guarantees that, if $\Psi(t)$ is a solution of (6.50), then $\lambda\Psi(t)$ is also a solution representing the same physical state. When $\mathscr{H}$ has a bilinear form such as $\mathscr{H} = \Psi_l^* H_{ll'} \Psi_{l'}$, (6.50) reduces to the standard form, but he allowed terms nonbilinear and nonlinear in Ψ and Ψ^* in $\mathscr{H}$. However, his subtle proposal immediately risks spoiling the superposition principle and therefore cannot be advantageous.

We now turn to investigate the time derivative of the Schrödinger equations. The time-dependent Schrödinger equation was first proposed in the fourth communication of Schrödinger's series on wave mechanics (Schrödinger, 1926). He had no ideas as to how to check its validity. (For the time-independent Schrödinger equation, the reliable partner to which it could be referred for comparison was the Heisenberg matrix equation.) Schrödinger employed the form $i\hbar(\partial\Psi/\partial t)$ by referring to the linear (classical) wave equation:

$$\frac{\partial}{\partial t}\Psi = -\frac{i}{\hbar}E\Psi, \tag{6.51}$$

which, combined with the time-independent equation $H\Psi = E\Psi$, resulted in the familiar time-dependent Schrödinger equation. The form in (6.51) may be reasonable because the resultant time-dependent Schrödinger equation recovers the Hamilton–Jacobi equation as $\hbar \to 0$. The first-order time derivative in (6.51) manifests the novel complex nature of Ψ. In the context of quantum chaos, in particular, the phase of Ψ has been shown to embody gauge structure around avoided level crossings and diabolical conical intersections of hyper-energy surfaces (see sections 6.1–6.2).

The form (6.51), however, exhibits only a time-periodic wave. So long as Schrödinger chose (6.51) as the foundation for his thinking, the time-dependent Schrödinger equation inevitably excluded any temporal chaos. There is no decisive reason for us to cling to a time-periodic equation like (6.51) in constructing quantum mechanics. One may conceive several other variants of (6.51) which will still possess the linear features required to ensure the superposition principle. One promising possibility is to discretize time t and replace the time derivative in (6.51) by time difference, thus

$$i\hbar\frac{\partial}{\partial t}\Psi \to i\hbar(\Psi(t_{n+1}) - \Psi(t_n))/\Delta t \tag{6.52}$$

with Δt a suitable 'time quantum'. Then we shall have a discrete version of the time-dependent Schrödinger equation given by

$$\Psi(t_{n+1}) = U\Psi(t_n) \tag{6.53a}$$

with

$$U = 1 + H\,\Delta t/(i\hbar). \tag{6.53b}$$

Let us suppose that (6.53) is employed as a fundamental equation. This just means that the right-hand side of (6.53*b*) cannot be exponentiated and that U in (6.53) is no longer unitary. Using a set of time-independent orthonormal bases $\{|\phi_l\rangle\}$, $\Psi(t_n)$ can be expressed as

$$\Psi(t_n) = \sum_{l=1}^{D} c_l(t_n)|\phi_l\rangle, \tag{6.54}$$

where D may be either finite or infinite and $\{c_l\}$ are complex coefficients. Equation (6.53) thereby becomes a linear discrete map for column vectors $\mathbf{c}(t_n) = (c_1(t_n), c_2(t_n), \ldots, c_D(t_n))$:

$$\mathbf{c}(t_{n+1}) = \hat{U}\mathbf{c}(t_n) \tag{6.55a}$$

with

$$\hat{U}_{ll'} = \langle l|U|l'\rangle. \tag{6.55b}$$

To recover the unitarity (i.e., conservation law for probability amplitude $|\Psi|^2$), we must impose a constraint on the map (6.55). Noting the normalization for state vectors

$$|\mathbf{c}(t_n)|^2 = \sum_{l=1}^{D} |c_l(t_n)|^2 = 1, \tag{6.56}$$

let us consider an ensemble of initial state vectors. This ensemble constitutes the unit hypersphere. By each application of the map in (6.55), this object will be stretched, but should simultaneously be folded so as to accommodate the original geometry, i.e., unit hypersphere. If actual experiments were carried out at time measured in units Δt, we could confirm unitarity at all these discrete times. The map thus constructed is area-preserving in the sense that the total surface area of the unit hypersphere is conserved.

To illustrate my speculation above, let us take Arnold's cat map (Arnold and Avez, 1968):

$$\begin{bmatrix} x_{n+1} \\ y_{n+1} \end{bmatrix} = \begin{bmatrix} 1 & 1 \\ 1 & 2 \end{bmatrix} \begin{bmatrix} x_n \\ y_n \end{bmatrix} \quad (\text{mod } 1). \tag{6.57}$$

Equation (6.57) is an amazing linear and area-preserving map and, like other low-dimensional linear maps such as the Bernoulli shift and Baker's transformation, has a mixing property. Equation (6.57) can be interpreted as a map of the unit sphere onto itself if variables x, y are replaced by polar θ and azimuthal ϕ angles as $x \to \theta/\pi$; $y \to \phi/2\pi$. Arnold's cat pasted on the sphere will evolve as shown in fig. 6.3.

In this example, radial vectors have real components and cannot be substitutes for state vectors. Furthermore, in contrast to the fully chaotic map (C system) as above, the linear map (6.55) with a folding mechanism possesses the following generic features: (i) the matrix $\hat{U}$ in (6.55) depends significantly on the nature of the Hamiltonian H; (ii) periodic orbits and KAM tori may coexist with chaos; (iii) the degree of folding decreases with decreasing Δt. In particular, no folding is necessary in the limit $\Delta t \to 0$ when the unitarity of $\hat{U}$ in (6.55) is satisfied. Careful consideration will indeed be demanded in writing down an expression for the folding mechanism associated with map (6.55), but my idea indicates that, instead of referring exclusively to time-periodic waves in (6.51), one may have another choice, namely of referring to time-chaotic waves.

Let us proceed to several results of my argument based on the discrete map (6.55). (i) So long as time-periodic evolutions of Ψ with period h/E

Fig. 6.3 Arnold cat pasted on unit sphere.

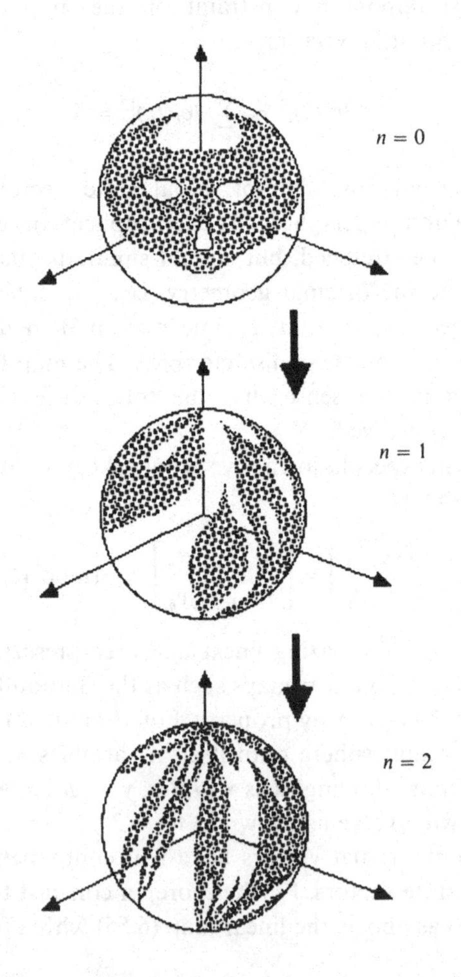

are concerned, distinct stationary states are available and, by suppressing the time-dependent factor $\exp(-iEt/\hbar)$, the ordinary time-independent Schrödinger equation will be recovered. (ii) In the semiclassical limit, if the limit $\Delta t \to 0$ is taken before the limit $\hbar \to 0$, the folding mechanism will become inactive and time difference reverts to time derivative in (6.52). Eventually the Hamilton–Jacobi equation can be recovered in the classical limit. (iii) When Ψ shows temporal chaos, the most interesting result will be found: instead of discrete energy levels, we have continuum or singular continuous spectra even for nondriven systems. Though stationary states can be defined, the adiabatic ansatz can no longer be fulfilled. Hence we have the possibility of establishing a one-to-one correspondence between classical and quantum chaos. (iv) Time discretization will necessitate much more complete discretization of space–time.

For the purpose of accommodating temporal chaos of wavefunctions Ψ, I have just proposed a way of improving Schrödinger's formalism of quantum mechanics. It should be remarked, however, that for unbounded and open systems the ordinary time-dependent equation can have solutions Ψ exhibiting ergodic, mixing and Bernoulli features. In these systems, D (dimensionality of Hilbert space) $= \infty$. Using the expansion of Ψ in (6.54), $i\hbar\dot{\Psi} = H\Psi$ can be reduced to the equation for an infinite harmonic solid or its modification for hypothetical particles located at $\{c_l(t)\}$ (see (6.54)), whose global Kolmogorov–Sinai entropy is positive (Lanford and Lebowitz, 1975; Goldstein *et al.*, 1975) as discussed in chapter 5. Here again we have continuum spectra. In any event, if temporal chaos is brought into quantum mechanics, our ideas on time, stationary states, the adiabatic ansatz and their relationship with the uncertainty principle should be carefully re-examined. These intriguing questions (even within nonrelativistic quantum mechanics) would be comparable to Einstein's great challenge in 1905 which revolutionized Galilean thought on space and time.

Any idea as to how to make quantum mechanics nonlinear should of course be tested by an accumulation of experiments on both microscopic and mesoscopic systems, e.g., conductance fluctuations in mesoscopic semiconductor devices, anomalous oscillations in magnetic resonance, ionization of hydrogen atoms by microwaves, molecular dissociations, etc.

In nature, external sources of randomness such as noise, impurities and random potentials are inevitable elements. These effects have so far been regarded as the unique source of spectral width in magnetic resonance, fluctuations in conductance, etc., but quantum chaos – deterministic randomness – may be a novel alternative candidate for spectral width and fluctuations in various observables. Quantum chaos here is not merely a fingerprint of chaos but may be a new paradigm of nonlinear dynamics that deserves much more revolutionary insights. One may try, in speculation, to go beyond Schrödinger's or Feynman's framework of quantum mechanics. I hope that readers will share with us these dramatic and exciting developments as we approach a new century.

References

Ablowitz, M. J., Ramani, A. and Segur, H. (1980). *J. Math. Phys.* **21**, 715.

Abrahams, E., Anderson, P. W., Licciardello, D. C., and Ramakrishnan, T. V. (1979). *Phys. Rev. Lett.* **42**, 673.

de Aguiar, F. M. and Rezende, S. M. (1986). *Phys. Rev. Lett.* **56**, 1070.

Anderson, P. W. (1973). *Mater. Res. Bull.* **8**, 153.

Anderson, P. W. (1981). In *Order and Fluctuations in Equilibrium and Nonequilibrium Statistical Mechanics*, ed. G. Nicolis, G. Dewel and J. W. Turner. New York: Wiley Interscience.

Anderson, P. W. (1987). *Science* **235**, 1196

Arnold, V. I. (1978). *Mathematical Methods of Classical Mechanics.* New York: Springer.

Arnold, V. I. and Avez, A. (1968). *Ergodic Problems of Classical Mechanics.* New York: Benjamin.

Ashcroft, N. W. and Mermin, N. D. (1976). *Solid State Physics.* Ithaca: Cornell University Press.

Aurich, R., Sieber, M. and Steiner, F. (1988). *Phys. Rev. Lett.* **61**, 483.

Avishai, Y. and Band, Y. B. (1987). *Phys. Rev. Lett.* **58**, 2251.

Avishai, Y. and Band, Y. B. (1989). *Phys. Rev. Lett.* **62**, 2527.

Balazs, N. L. and Voros, A. (1989). *Ann. Phys. (N.Y.)* **190**, 1.

Balian, R. and Bloch, C. (1972). *Ann. Phys.* **69**, 76.

Balian, R. and Bloch, C. (1974). *Ann. Phys.* **85**, 514.

Baranger, H. U., DiVincenzo, D. P., Jalabert, R. A. and Stone, A. D. (1991). *Phys. Rev.* **B44**, 10637.

Bayfield, J. E. and Koch, P. M. (1974). *Phys. Rev. Lett.* **33**, 258.

Beenakker, C. W. J. and van Houten, H. (1991). In *Solid State Physics: Advances in Research and Applications*, ed. H. Ehrenreich and D. Turnbull. New York: Academic.

Bell, J. S. (1964). *Physics* **1**, 195.

Benettin, G. and Strelcyn, J. M. (1978). *Phys. Rev.* **17A**, 773.

Berry, M. V. (1981). *Ann. Phys. (N.Y.)* **131**, 163.

Berry, M. V. (1984). *Proc. R. Soc. London* **A392**, 45.

Berry, M. V. (1985). *Proc. R. Soc. London* **A400**, 229.

Berry, M. V. (1986). In *Quantum Chaos and Statistical Nuclear Physics*, ed. T. Seligman and H. Nishioka. Berlin: Springer.

Berry, M. V. (1990). *Proc. R. Soc. London* **A430**, 405.

Berry, M. V. and Balazs, N. L. (1979). *J. Phys.* **A12**, 625.

Berry, M. V. and Keating, J. P. (1990). *J. Phys.* **A23**. 4839.

Berry, M. V. and Tabor, M. (1977a). *J. Phys.* **A10**, 371.

Berry, M. V. and Tabor, M. (1977b). *Proc. R. Soc. London* **A356**, 375.

Birkhoff, G. D. (1931). *Proc. Natl Acad. Sci. USA.* **17**, 656.

Bloembergen, N. and Wang, S. (1954). *Phys. Rev.* **93**, 72.

Bocchieri, P. and Loinger, A. (1957). *Phys. Rev.* **107**, 337.

Bocchieri, P. and Loinger, A. (1958). *Phys. Rev.* **111**, 668.

Bocchieri, P. and Loinger, A. (1959). *Phys. Rev.* **114**, 948.

Bogomolny, E. B. (1988). *Physica* **D31**, 169.

Bohigas, O. and Giannoni, M. J. (1984). In *Mathematical and Computational Methods in Nuclear Physics*, ed. J. S. Dehesa, J. M. G. Gomez and A. Polls. Berlin: Springer.

Bohigas, O., Giannoni, M. J. and Schmit, C. (1984). *Phys. Rev. Lett.* **52**, 1.

Bohr, A. and Mottelson, B. R. (1953). *K. Dan. Vidensk. Selsk. Mat. Fys. Medd.* **27**, 1.

Bohr, N. (1911). Dissertation, Copenhagen.

Bowen, R. (1975). *Equilibrium States and the Ergodic Theory of Anosov Diffeomorphisms*. Berlin: Springer.

Brody, T. A., Flores, J., French, J. B., Mello, P. A., Pandey, A. and Wong, S. S. M. (1981). *Rev. Mod. Phys.* **53**, 385.

Bryant, P., Jeffries, C. and Nakamura, K. (1988a). *Phys. Rev. Lett.* **60**, 1185.

Bryant, P., Jeffries, C. and Nakamura, K. (1988b). *Phys. Rev.* **A38**, 4223.

Bunimovich, L. A. (1974). *Funct. Anal. Appl.* **8**, 254.

Carroll, T. L., Pecora, L. M. and Rachford, F. J. (1987). *Phys. Rev. Lett.* **59**, 2891.

Casati, G., Chirikov, B. V., Izrailev, F. M. and Ford, J. (1979). In *Stochastic Behavior in Classical and Quantum Hamiltonian Systems*, ed. G. Casati and J. Ford. Berlin: Springer.

Cascon, A., Koiller, J. and Rezende, S. M. (1991). *Physica* **D54**, 98.

Child, M. S. (1974). *Molecular Collision Theory*. New York: Academic.

Chirikov, B. V. (1979). *Phys. Rep.* **52**, 263.

Cvitanović, P. and Eckhardt, B. (1989). *Phys. Rev. Lett.* **63**, 829.

Damon, R. W. (1953). *Rev. Mod. Phys.* **25**, 239.

Daniel, M., Kruskal, M. D., Lakshmanan, M. and Nakamura, K. (1992). *J. Math. Phys.* **33**, 771.

Darboux, G. (1896). *Leçons sur la théorie générale des surfaces*, vol. 4. Paris: Gauthier-Villars.

Demkov, Y. N. and Osherov, V. I. (1967). *Zh. Eksp. Theor. Fiz.* **53**, 1589.

Dyson, F. J. (1962a). *J. Math. Phys.* **3**, 140.

Dyson, F. J. (1962b). *J. Math. Phys.* **3**, 157.

Dyson, F. J. (1962c). *J. Math. Phys.* **3**, 1191.

Dyson, F. J. (1962d). *J. Math. Phys.* **3**, 1199.

Dyson, F. J. (1972). *J. Math. Phys.* **13**, 90.

Ehrenfest, P. (1916). *Ann. Phys. (Leipzig)* **51**, 327.

Einstein, A. (1917). *Verh. Dtsch. Phys. Ges.* **19**, 82.

Einstein, A., Podolsky, B. and Rosen, N. (1935). *Phys. Rev.* **47**, 777.

Faddeev, L. D. and Takhtajan, L. A. (1987). *Hamiltonian Methods in the Theory of Solitons*. Berlin: Springer.

Fano, U. (1961). *Phys. Rev.* **124**, 1866.

Feynman, R. P. and Hibbs, A. R. (1965). *Quantum Mechanics and Path Integrals*. New York: McGraw-Hill.

Fishman, S., Grempel, D. R. and Prange, R. E. (1982). *Phys. Rev. Lett.* **49**, 509.

Frahm, H. and Mikeska, H. J. (1986). *Z. Phys.* **B65**, 249.

Gaspard, P., Rice, S. A. and Nakamura, K. (1989). *Phys. Rev. Lett.* **63**, 930.

Gaspard, P., Rice, S. A., Mikeska, H. J. and Nakamura, K. (1990). *Phys. Rev.* **A42**, 4015.

Gibbons, J. and Hermsen, T. (1984). *Physica* **D11**, 337.

Gibson, G. and Jeffries, C. (1984). *Phys. Rev.* **A29**, 811.

Goldstein, S., Lebowitz, J. L. and Aizenman, M. (1975). In *Dynamical Systems, Theory and Applications*, ed. J. Moser. Berlin: Springer.

Greene, J. M. (1979). *J. Math. Phys.* **20**, 1183.

Greene, J. M., MacKay, R. S., Vivaldi, F. and Feigenbaum, M. J. (1981). *Physica* **3D**, 468.

Guckenheimer, J. and Holmes, P. (1983). *Nonlinear Oscillations, Dynamical Systems and Bifurcations of Vector Fields.* New York: Springer.

Gutzwiller, M. C. (1967). *J. Math. Phys.* **8**, 1979.

Gutzwiller, M. C. (1971). *J. Math. Phys.* **12**, 343.

Gutzwiller, M. C. (1982). *Physica* **D5**, 183.

Gutzwiller, M. C. (1990). *Chaos in Classical and Quantum Mechanics.* Berlin: Springer.

Gutzwiller, M. C. and Mandelbrot, B. B. (1988). *Phys. Rev. Lett.* **60**, 673.

Haake, F., Kus, M. and Scharf, R. (1987). *Z. Phys.* **B65**, 381.

Halsey, T. C., Jensen, M. H., Kadanoff, L. P., Procaccia, I. and Shraiman, B. I. (1986). *Phys. Rev.* **A33**, 1141.

Hartwick, T. S., Peressini, E. R. and Weiss, M. T. (1961). *J. Appl. Phys.* **32**, 223S.

Heller, E. J. (1984). *Phys. Rev. Lett.* **53**, 1515.

Hogg, T. and Huberman, B. A. (1982). *Phys. Rev. Lett.* **48**, 711.

Husimi, K. (1940). *Proc. Phys. Math. Soc. Jpn.* **22**, 264.

Ishio, H. and Nakamura, K. (1992). *Phys. Rev.* **A46**, R2193.

Iu, C., Welch, G. R., Kash, M. M., Hsu, L. and Kleppner, D. (1989). *Phys. Rev. Lett.* **63**, 1133.

Jackiw, R. (1988). *Comments At. Mol. Phys.* **21**, 71.

Jalabert, R. A., Baranger, H. U. and Stone, A. D. (1990). *Phys. Rev. Lett.* **65**, 2442.

Keller, J. B. (1958). *Ann. Phys. (N.Y.)* **4**, 180.

Keller, J. B. and Rubinow, S. (1960). *Ann. Phys. (N.Y.)* **9**, 27.

Klauder, J. R. and Skagerstam, B. (1985). *Coherent States.* Singapore: World Scientific.

Kleppner, D., Littman, M. C. and Zimmerman, M. L. (1983). In *Rydberg States of Atoms and Molecules*, ed. R. F. Stebbings and F. B. Dunning. Cambridge: Cambridge University Press.

Korsch, H. J. and Berry, M. V. (1981). *Physica* **3D**, 627.

Kuratsuji, H. and Iida, S. (1988). *Phys. Rev.* **D37**, 441.

Lakshmanan, M. and Nakamura, K. (1984). *Phys. Rev. Lett.* **53**, 2497.

Landau, L. D. (1930). *Z. Phys.* **64**, 629.

Landau, L. D. (1932). *Phys. Z. Sowjetunion.* **2**, 46.

Lanford, O. E. and Lebowitz, J. L. (1975). *Lecture Notes in Physics*, vol. 38. Berlin: Springer.

Laughlin, R. B. (1987). *Nucl. Phys.* **B** (Proc. Suppl.) **3**, 213.

Lazutkin, V. F. (1973). *Izv. Akad. Nauk. Ser. Math.* **37**, 186.

Leff, H. S. (1964). *J. Math. Phys.* **5**, 763.

Leopold, J. C. and Percival, I. C. (1978). *Phys. Rev. Lett.* **41**, 944.

Lichtenberg, A. J. and Lieberman, M. A. (1983). *Regular and Stochastic Motion.* New York: Springer.

Likharev, K. K. and Zorin, A. B. (1985). *J. Low Temp. Phys.* **59**, 347.

Longuet-Higgins, H. C., Öpik, U., Pryce, M. H. L. and Sack, R. A. (1958). *Proc. R. Soc. London* **A244**, 1.

Magyari, E., Thomas, H., Weber, R., Kaufman, C. and Müller, G. (1987). *Z. Phys.* **B65**, 363.

Mandelbrot, B. B. (1982). *The Fractal Geometry of Nature.* New York: Freeman.

McDonald, S. W. and Kaufman, A. N. (1979). *Phys. Rev. Lett.* **42**, 1189.

Mehta, M. L. (1967). *Random Matrices and the Statistical Theory of Energy Levels.* New York: Academic.

Mehta, M. L. and Dyson, F. J. (1963). *J. Math. Phys.* **4**, 713.

Mehta, M. L. and Gaudin, M. (1960). *Nucl. Phys.* **18**, 420.

Mehta, M. L. and Gaudin, M. (1961). *Nucl. Phys.* **22**, 340.

Mino, M. and Yamazaki, H. (1986). *J. Phys. Soc. Jpn.* **55**, 4168.

Mino, M., Yamazaki, H. and Nakamura, K. (1989). *Phys. Rev.* **B40**, 5279.

Morgenthaler, F. M. (1960). *J. Appl. Phys.* **31**, 95S.

Mott, N. F. and Jones, H. (1936). *The Theory of the Properties of Metals and Alloys.* Oxford: Oxford University Press.

Nakamura, K. (1989). *Prog. Theor. Phys. Suppl.* **98**, 383.

Nakamura, K. (1990). In *Nonlinear Evolution Equations: Integrability and Spectral Methods*, ed. A. Degasperis, A. P. Fordy and M. Lakshmanan. Manchester: Manchester University Press.

Nakamura, K. and Bishop, A. R. (1986). *Phys. Rev.* **B33**, 1963.

Nakamura, K., Bishop, A. R. and Shudo, A. (1989). *Phys. Rev.* **B39**, 12422.

Nakamura, K. and Ishio, H. (1992). *J. Phys. Soc. Jpn.* **61**, 3939.

Nakamura, K. and Lakshmanan, M. (1986). *Phys. Rev. Lett.* **57**, 1661.

Nakamura, K., Lakshmanan, M., Gaspard, P. and Rice, S. A. (1992). *Phys. Rev.* **A46**, 6311.

Nakamura, K. and Mikeska, H. J. (1987). *Phys. Rev.* **A35**, 5294.

Nakamura, K., Nakahara, Y. and Bishop, A. R. (1985). *Phys. Rev. Lett.* **54**, 861.

Nakamura, K., Ohta, S. and Kawasaki, K. (1982). *J. Phys.* **C15**, L143.

Nakamura, K., Okazaki, Y. and Bishop, A. R. (1986). *Phys. Rev. Lett.* **57**, 5.

Nakamura, K. and Sasada, T. (1982). *J. Phys.* **C15**, L915.

Nakamura, K. and Thomas, H. (1988). *Phys. Rev. Lett.* **61**, 247.

von Neumann, J. (1929). *Z. Phys.* **57**, 30.

von Neumann, J. and Wigner, E. P. (1929). *Phys. Z.* **30**, 467.

Odlyzko, A. M. (1987). *Math. Comp.* **48**, 273.

Ohta, S. and Nakamura, K. (1983). *J. Phys.* **C16**, L605.

Olshanetsky, M. A. and Perelomov, A. M. (1981). *Phys. Rep.* **71**, 313.

Pechukas, P. (1983). *Phys. Rev. Lett.* **51**, 943.

Pechukas, P. (1984). *J. Phys. Chem.* **88**, 4823.

Peierls, R. (1933a). *Z. Phys.* **80**, 763.

Peierls, R. (1933b). *Z. Phys.* **81**, 186.

Peierls, R. (1979). *Surprises in Theoretical Physics*. Princeton: Princeton University Press.

Percival, I. C. (1977). *Adv. Chem. Phys.* **36**, 1.

Pietronero, L. and Siebesma, A. P. (1986). *Phys. Rev. Lett.* **57**, 1098.

Pomphrey, N. (1974). *J. Phys.* **B7**, 1909.

Radcliffe, J. M. (1971). *J. Phys.* **A4**, 313.

Rice, S. A., Gaspard, P. and Nakamura, K. (1992). In *Advances in Classical Trajectory Methods Vol. 1*, ed. W. L. Hase. Connecticut: JAI Press.

Robnik, M. and Berry, M. V. (1985). *J. Phys.* **A18**, 1361.

Roukes, M. L. and Alerhand, O. L. (1990). *Phys. Rev. Lett.* **65**, 1651.

Ruelle, D. (1978). *Thermodynamic Formalism.* Massachusetts: Addison-Wesley.

Ruelle, D. and Takens, F. (1971). *Commun. Math. Phys.* **20**, 167.

Schlömann, E., Green, J. J. and Milano, U. (1960). *J. Appl. Phys.* **31**, 386S.

Schrödinger, E. (1926). *Ann. Phys. (Leipzig)* **81**, 109.

Selberg, A. (1956). *J. Indian Math. Soc.* **20**, 47.

Shapere, A. and Wilczek, F. (1989). *Geometric Phases in Physics.* Singapore: World Scientific.

Shepelyansky, D. L. (1986). *Phys. Rev. Lett.* **56**, 677.

Sinai, Y. G. (1976). *Introduction to Ergodic Theory.* Princeton: Princeton University Press.

Sinai, Y. G. and Volkoviski, K. L. (1977). *Funct. Anal. Appl.* **5**, 19.

Sklyanin, E. K. (1979). *Dokl. Akad. Nauk.* **244–245**, 1337.

Stein, D. L. (1980). *J. Chem. Phys.* **72**, 2869.

Stueckelberg, E. C. G. (1932). *Helv. Phys. Acta* **5**, 369.

Suhl, H. (1957). *J. Phys. Chem. Solids* **1**, 209.

Sutherland, B. (1971). *Phys. Rev.* **A4**, 2019.

Takahashi, K. and Saito, N. (1985). *Phys. Rev. Lett.* **55**, 645.

Takami, T. and Hasegawa, H. (1992). *Phys. Rev. Lett.* **68**, 419.

Tanner, G., Scherer, P., Bogomonly, E. B., Eckhardt, B. and Wintgen, D. (1991). *Phys. Rev. Lett.* **67**, 2410.

Teller, E. (1931). *Z. Phys.* **67**, 311.

Thouless, D. J. (1960). *Phys. Rev.* **117**, 1256.

van Leeuwen, H. J. (1921). *J. Phys. (Paris)* **2**, 361.

Van Vleck, J. H. (1928). *Proc. Natl Acad. Sci. USA* **14**, 178.

Waldner, F., Barberis, D. R. and Yamazaki, H. (1985). *Phys. Rev.* **A31**, 420.

Weinberg, S. (1989). *Ann. Phys. (N.Y.)* **194**, 336.

Weissman, Y. and Jortner, J. (1982). *J. Chem. Phys.* **77**, 1486.

Wen, X. G., Wilczek, F. and Zee, A. (1989). *Phys. Rev.* **B39**, 11413.

Weyl, H. (1927). *Z. Phys.* **46**, 1.

Wheeler, J. A. and Zurek, W. H. (1983). *Quantum Theory of Measurement*. Princeton: Princeton University Press.

Wiese, G. and Benner, H. (1990). *Z. Phys.* **B79**, 119.

Wigner, E. (1932). *Phys. Rev.* **40**, 749.

Wintgen, D. (1988). *Phys. Rev. Lett.* **61**, 1803.

Yamazaki, H. and Warden, M. (1986). *J. Phys. Soc. Jpn.* **55**, 4477.

Yukawa, T. (1985). *Phys. Rev. Lett.* **54**, 1883.

Zakharov, V. E., L'vov, V. S. and Starobinets, S. S. (1974). *Usp. Fiz. Nauk* **114**, 609.

Zaslavsky, G. M. (1981). *Phys. Rep.* **80**, 157.

Zener, C. (1932). *Proc. R. Soc. London* **A137**, 696.

Ziglin, S. L. (1983a). *Funct. Anal. Appl.* **16**, 181.

Ziglin, S. L. (1983b). *Funct. Anal. Appl.* **17**, 6.

Index

Printed in the United States
By Bookmasters